SpringerBriefs in Food, Health, and Nutrition

Springer Briefs in Food, Health, and Nutrition present concise summaries of cutting edge research and practical applications across a wide range of topics related to the field of food science.

For further volumes:
http://www.springer.com/series/10203

Taylor C. Wallace · Douglas MacKay
Rend Al-Mondhiry · Haiuyen Nguyen
James C. Griffiths

Dietary Supplement Regulation in the United States

Springer

Taylor C. Wallace
Douglas MacKay
Rend Al-Mondhiry
Haiuyen Nguyen
James C. Griffiths
Council for Responsible Nutrition
Washington, DC, USA

ISBN 978-3-319-01501-9 ISBN 978-3-319-01502-6 (eBook)
DOI 10.1007/978-3-319-01502-6
Springer Cham Heidelberg New York Dordrecht London

Library of Congress Control Number: 2013946031

Printed on acid-free paper

Springer is part of Springer Science+Business Media (www.springer.com)

Contents

Abstract

Dietary supplements play an important role in the growing nutrition industry in the United States and across the globe. Health and food safety authorities such as Health Canada, the European Commission, and the US Food and Drug Administration continue to develop and implement regulations aimed at ensuring consumer access to safe and effective products. In the United States, the Federal Food, Drug, and Cosmetic Act (FDCA) governs dietary supplements under the larger category of food. However, it was not until 1994 that the Dietary Supplement Health and Education Act amended the FDCA to give definition to the term "dietary supplement" and to establish a regulatory framework specific to supplements. This Springer Brief aims to review and discuss the many current statutes and regulations addressing dietary ingredients, manufacturing standards, safety, labeling, and claims, all of which promote the safe use of dietary supplements and preserve continued consumer access. This Brief also provides an overview of how dietary supplements are regulated in Europe, Canada, and Japan.

Chapter 1
Dietary Supplement Regulation in the United States

1.1 Introduction

Dietary supplements, also called food or health supplements in Europe and Asia, are a growing category of consumer products across the globe. Health and food safety authorities such as Health Canada, the European Commission, and the U.S. Food and Drug Administration (FDA) continue to develop and implement regulations aimed at ensuring consumer access to safe and effective products. In the United States, the Federal Food, Drug, and Cosmetic Act (FDCA) [1] governs a wide variety of products, including dietary supplements, which are regulated as a category of food. However, it was not until 1994 that the United States Congress enacted the Dietary Supplement Health and Education Act (DSHEA) [2], which amended the FDCA and established specific provisions for dietary supplements.

Dietary supplements, a term which initially referred only to vitamins and minerals, have been regulated in the United States for over a century, beginning with the Pure Food and Drug Act (PFDA) of 1906 [3] and followed by the 1938 FDCA [4]. The latter established a category of "foods for special dietary uses" to include vitamin supplements, fortified foods, and infant formula. Many years later, the 1976 amendments to the FDCA clarified that a "special dietary use" could include "supplying a vitamin, mineral or other ingredient for use by man to supplement his diet by increasing the total dietary intake." [5] DSHEA expanded this definition by defining the term *dietary supplement*. Most recently, Congress enacted the Food Safety Modernization Act of 2010 (FSMA) [6]. This sweeping law represents a major shift in the regulation of foods in the United States and authorized FDA to develop regulations affecting all aspects of food production and distribution with a focus on prevention of food-borne illness. Currently, the FDA is in the process of implementing the different elements of FSMA, including proposed rulemaking that addresses good manufacturing practices (GMPs) for foods (including dietary ingredients) and verification of foreign suppliers [7, 8]. Although FSMA's mandates focus on conventional foods, some provisions also impact dietary supplements.

T.C. Wallace et al., *Dietary Supplement Regulation in the United States*,
SpringerBriefs in Food, Health, and Nutrition,
DOI 10.1007/978-3-319-01502-6_1,

This Springer Brief will review and discuss the applicable regulations for dietary supplements pertaining to manufacturing standards, safety, labeling, and claims. In addition, an overview of dietary supplement regulation across the globe provides a glimpse of the variety of approaches taken by governments to provide safe consumer access to these products.

1.1.1 A Historical Overview of Legislation and Regulation Before the 1994 Dietary Supplement Health and Education Act

The first dietary supplements became available in the 1920s, and consisted of common food constituents and nutrients. Cod liver oil, used to enhance vitamin A and D intake, was one of the pioneer products available for purchase [8]. In 1934, Nutrilite Company marketed the first multivitamin–multimineral tablet by drying and compressing vegetable and fruit juice concentrates [9].

The PFDA of 1906 defined for the first time in federal statute the terms *misbranding* and *adulteration* and provided penalties for each, thereby paving the way for new safety standards for food. In 1938, the FDCA [4] defined "tolerances" (maximum allowed levels) for unavoidable harmful substances (adulterants), and set identity and quality standards for food. Section 402 of the FDCA [1] states that a food is declared unsafe or adulterated if it contains a substance "which may render it injurious to health." As a category of food, dietary supplements are also subject to this provision. Section 403 of the FDCA [1] refers to foods for special dietary uses, including dietary supplements, which FDA may deem misbranded unless the "label bears such information concerning its vitamin, mineral, and other dietary properties as the Secretary of Health and Human Services determines to be, and by regulations prescribes as, necessary in order to fully inform purchasers as to its value for such uses."

Subsequently, FDA implemented regulations that pertain specifically to dietary supplements and address nutrition labeling and content. In 1941, the FDA required foods for special dietary uses (i.e., foods with added vitamins or minerals) to have labeling declaring the name, quantity, and percent Minimum Daily Requirement of each added nutrient [10]. In the 1960s and 1970s, FDA attempted to limit products to approximately 150 % of the reference value for vitamins and minerals; otherwise, these products would be classified as drugs. The agency also tried to restrict the number of combinations of vitamins and minerals that could be sold by issuing a regulation in 1973 that established a Standard of Identity for vitamin and mineral supplements [11]. This rule was challenged and overturned on two separate occasions. In 1976, Congress passed the Proxmire Amendments (21 USC §350) [12], which prevented FDA from establishing a standard limiting vitamins and minerals in nutritional supplements and nullified the agency's authority to classify a vitamin or mineral product as a drug based solely on its potency. Rather, dietary supplements, and all foods, may be considered drugs if their labeling suggests intended use as drugs, which are defined as an article "intended for use in the diagnosis, cure, mitigation, treatment, or prevention of disease in man or other animals." [1] For example, niacin can be legally marketed in a fortified cereal and as a dietary supplement to help meet normal nutrient needs, and it may also be marketed as a drug to lower cholesterol levels.

In the 1990s, FDA renewed its efforts to limit access to dietary supplements with the publication of an advanced notice of proposed rulemaking (ANPR). The 1993 ANPR suggested that vitamins and minerals should have limited doses, amino acids are food additives not legal in supplements, and herbal products are inherently therapeutic and should not be sold as dietary supplements [13]. The possibility of overly restrictive regulations led to the drafting of DSHEA in 1994 by Senators Orrin Hatch and Tom Harkin. Table 1.1 provides a summary of relevant federal statutes/amendments and their mandates regarding dietary supplements since the enactment of the PFDA in 1906.

1.1.2 Dietary Supplement Health and Education Act of 1994

When DSHEA was enacted, approximately 4,000 dietary supplement products were available on the market [14]. Legislators, recognizing the benefits of nutritional supplements and the integral role of the supplement industry in the U.S. economy, passed DSHEA with two primary goals: to provide consumers access to a variety of dietary supplements and to provide more information to consumers about the intended use of supplement products. DSHEA accomplished these goals by defining dietary supplements, affirming food safety standards, authorizing GMP regulations, and addressing nutritional labeling and claims.

1.1.2.1 Defining Dietary Supplements

Dietary supplements are regulated as a category of foods, and thus are subject to the basic legal and regulatory requirements applicable to conventional foods under the FDCA, except in cases in which they are preempted by DSHEA or other relevant statutes. DSHEA defines the term *dietary supplement* and establishes a regulatory framework for dietary supplements. While providing consumers broad access to dietary supplements, DSHEA also provides FDA with authority to remove from the market products that pose a "significant or unreasonable" risk to consumers. The law also requires dietary supplement products to be accurately labeled to inform consumers about their intended use.

Under DSHEA, a dietary supplement is defined as:

> … [a] product that is intended to supplement the diet that may contain one or more dietary ingredients. A dietary ingredient may be: a vitamin or a mineral, an herb or other botanical, an amino acid, a dietary substance for use by man to supplement the diet by increasing the total dietary intake, or a concentrate, metabolite, constituent, extract, or combination of the preceding ingredients, and, that meet other criteria specified in section 201(ff)(2)-(3).

Table 1.1 Summary of federal statutes/amendments and their mandates on dietary supplements

Year	Statute/amendment	Citation
1906	Pure Food and Drug Act (PFDA): Adulteration standard that prohibits any added poisonous or deleterious substance injurious to health in food	59th Cong. Sess. 1. Chp. 3915, p. 768–772; cited as 34 U.S. Stats. 768
1938	Federal Food, Drug, and Cosmetic Act (FDCA): Authorizes foods to bear claims describing effects on [normal] structure or function of the body; establishes a category of foods for special dietary uses; authorizes the U.S. Food and Drug Administration (FDA) to regulate such products, which will be deemed misbranded unless the label bears information concerning vitamin, mineral, or other dietary properties as prescribed by FDA regulations as necessary to fully inform purchasers as to the value of the food for such special dietary uses; grants FDA authority to inspect facilities	Public Law 75-717, 52 Stat. 1040
1938	Wheeler-Lea Act: Amends the Federal Trade Commission (FTC) Act to grant FTC advertising oversight of FDA-regulated products (except prescription drugs)	Public Law 75-447, 52 Stat. 111
1958	Food Additives Amendment: Establishes a premarket approval system for food additives through FDA petition process; defines a food additive as any substance added to food (directly or indirectly), unless the substance is generally recognized as safe for its intended use	Public Law 85-929, 72 Stat. 1784
1976	Proxmire Amendment: Prohibits FDA from classifying vitamin and mineral supplements as drugs based solely on their combinations or potency	Public Law 94-278
1990	Nutrition Labeling and Education Act (NLEA): Requires all food labels to contain specific information on the nutritional content (mandates the Nutrition Facts label) and authorizes FDA to consider and permit by regulation claims describing the relationship of specific nutrients to reduced risk of disease (i.e., health claims)	Public Law 101-535
1992	Dietary Supplement Act: Prohibits the implementation of NLEA with respect to dietary supplements except for the approved health claim provision, creating a moratorium to provide Congress and FDA time to draft DSHEA	Public Law 102-6571
1994	Dietary Supplement Health and Education Act (DSHEA): Defines the term *dietary supplement*; exempts dietary ingredients from the food additive provisions in the FDCA; establishes a new safety standard for dietary supplements; and authorizes FDA to impose requirements for good manufacturing practices	Public Law 103-417, 108 Stat. 4332

1996	Food Quality Protection Act: Amends the Federal Insecticide, Fungicide, and Rodenticide Act and the FDCA to require complete reassessment of all pesticide tolerances; mandates a single, scientifically based standard for all pesticide tolerances in all foods (including dietary supplements)	Public Law 104-170
1997	Food and Drug Administration Modernization Act (FDAMA): Permits the use of health claims and nutrient content claims based on authoritative statements by a scientific body of the U.S. government (e.g., National Institutes of Health) provided that premarket notification is sent to FDA	Public Law 105-115
2002	Public Health Security and Bioterrorism Preparedness and Response Act: Requires FDA registration of all food manufacturers and notification in advance of importation of food, including dietary supplements and raw materials	Public Law 107-188
2004	Anabolic Steroid Control Act: Prohibits steroid precursors to be sold in dietary supplements	Public Law 108-358
2004	Food Allergen Labeling and Consumer Protection Act: Requires disclosure on food and dietary supplement labels of 8 major allergens	Public Law 108-132, 118 Stat. 905
2006	Dietary Supplement and Nonprescription Drug Consumer Protection Act: Requires manufacturers and distributors to maintain records of all adverse event reports and to report all serious adverse event report data to FDA	Public Law 109-462
2007	Food and Drug Administration Amendments Act: Prohibits the introduction into interstate commerce of any food which contains an added drug	Public Law 110-185
2011	Food Safety Modernization Act (FSMA): Provides FDA with authority to issue a mandatory recall of any food product (except infant formula, which is already subject to FDA mandatory recall authority), including dietary supplements. Other provisions also apply	Public Law 111-353, 124 Stat. 3885

Note: Table 1.1 modified from Soller et al. [113] and reprinted with permission from the American Botanical Council

DSHEA also excluded dietary ingredients from the definition of food additives, which have a separate set of requirements. With some exceptions, ingredients used in dietary supplements prior to the enactment of DSHEA on October 15, 1994 are considered "grandfathered" and may continue to be used without any notification to FDA. A dietary ingredient introduced to the market after the passage of DSHEA is considered a new dietary ingredient (NDI). Manufacturers must submit a notification to the FDA at least 75 days before introducing a NDI to market to provide information regarding the identity of ingredient and justification that the NDI "will reasonably be expected to be safe" [2].

Other components added to dietary supplements, such as preservatives, binders, and encapsulation materials, are subject to the Food Additives Amendment of 1958 [15]; they must be FDA-approved food additives or generally recognized as safe (GRAS) substances. Similarly, color additives must be approved for use in food and dietary supplements and are regulated under the Color Additives Amendment of 1960 [16].

By definition, dietary supplements must be intended for ingestion [17]. Products that may be inhaled, absorbed through the skin, or injected through the epidermis cannot be legally marketed or sold as dietary supplements. The permissible dosage forms include tablets, capsules, powders, soft gels, or liquids [18]. Products intended as conventional foods (e.g., brownies, cereal, soda) may contain food additives, and/or GRAS substances, but cannot be marketed as dietary supplements. Energy drinks are sold in the market as both beverages (conventional food) and as liquid dietary supplements. In 2009, FDA published a draft guidance [19] for industry to clarify the differences between liquid conventional foods and dietary supplements; a final guidance is anticipated in 2013.

1.1.2.2 Developing Safety Standards

Dietary supplements are subject to the basic adulteration standard for foods. In addition, a dietary supplement is considered unsafe (adulterated) if it "presents a significant or unreasonable risk of illness or injury" under the conditions of use recommended or suggested on the product label or, in the absence of such recommendations or suggestions, under ordinary conditions of use.

Under DSHEA, the responsibility of ensuring the safety of a dietary supplement rests with the manufacturer. The law does not require supplement products to be registered with FDA or approved by the agency before being marketed. However, FDA has authority to remove any unsafe dietary supplement product from the market by demonstrating that a product is not reasonably expected to be safe. In 2004, FDA exercised this authority after determining that products containing ephedrine alkaloids were unsafe and called for their removal from the market [20].

1.1.2.3 Establishing Good Manufacturing Practices

DSHEA authorized FDA to develop GMP regulations for dietary supplements modeled after food current GMPs (cGMPs). FDA published an ANPR for dietary supplement cGMPs in 1997, and followed up with a proposed rule in 2003 and a final rule in 2007 [21]. The final rule, 21 CFR Part 111, requires all entities that manufacture, package, label, or hold dietary supplements to establish and follow cGMPs to ensure the quality and safety of finished products. The cGMP regulations mandate that dietary supplements meet the identity, purity, strength, and composition established in specifications and are properly packaged and labeled as specified in the master manufacturing record [21]. Raw material manufacturers that supply dietary and other ingredients are subject to food cGMPs found in 21 CFR Part 110.

1.1.2.4 Addressing Labeling and Claims

DSHEA also amended the FDCA, in part, by adding specific labeling requirements for dietary supplements and permitted optional label statements. A dietary supplement label must include a Supplement Facts panel, which is similar to the Nutrition Facts panel found on the label of conventional foods. The Supplement Facts panel must display the actual amounts of nutrients as well as their percent Daily Value (% DV). The final rule covering all aspects of labeling for dietary supplements was published by FDA in September 1997 [22].

DSHEA also defined permissible label claims for dietary supplements, including structure/function claims, which provide consumers with information about the intended use of a supplement. Structure/function claims have historically appeared on the labels of conventional foods and dietary supplements. However, DSHEA established regulatory procedures for such claims specific to dietary supplement labels. Structure/function claims describe "the role of a nutrient or dietary ingredient intended to affect normal structure or function in humans." A product may not claim to prevent or treat disease, which are claims permitted only for drugs. Because the supplement manufacturer is responsible for ensuring the accuracy and truthfulness of these claims, it must have substantiation for the structure/function claims it makes. Companies must also notify FDA within 30 days of making the structure/function claim, and products must bear a "disclaimer" stating that FDA has not evaluated the claim and that the product is not intended to "diagnose, treat, cure, or prevent any disease." FDA published a rule defining the scope of structure/function claims in January 2000 [23]. Prior to the passage of DSHEA, the Nutrition Labeling and Education Act of 1990 (NLEA) authorized FDA to allow "health claims" on food labeling [24]. Unlike structure/function claims, health claims describe a specific relationship between a food substance and a disease and must be evaluated and approved by FDA prior to use on a product label.

1.2 Current Good Manufacturing Practices

Collectively, cGMP regulations describe the required methods, systems, equipment, facilities, and controls for producing dietary supplements, food, medical devices, pharmaceutics, biologics, and veterinary products. Dietary supplement cGMPs have been in place since June 25, 2007, when FDA published the final rule,

Current Good Manufacturing Practice (cGMP) in Manufacturing, Packaging, Labeling, or Holding Operations for Dietary Supplements, in the *Federal Register* [21]. The final cGMP rule establishes minimum requirements necessary to ensure the efficacy, quality, purity, and safety of supplement products. All persons who import, export, manufacture, package, label, or hold a dietary supplement are required to adhere to cGMPs, and all dietary supplements available for use in the United States must be produced according to cGMPs.

After DSHEA authorized FDA to issue cGMP regulations for dietary supplements, in November 1995 representatives of the dietary supplement industry submitted to FDA a suggested outline for the cGMP regulations. FDA evaluated the outline and determined that it provided an "extremely useful starting point should FDA decide to proceed to rulemaking to adopt such regulations." The agency issued an ANPR in February 1997 [25], seeking stakeholder comments on the industry-submitted outline, among other issues. FDA issued the final cGMP rule in June 2007 [21]. The cGMP rule provided for a staggered compliance schedule based on company size to ensure that all impacted dietary supplement facilities had appropriate time to comply with the new rule. For large companies, defined as those with more than 500 employees, the final rule became effective in June 2008. Companies with 20–500 employees had until June 2009 to comply with the regulation, and very small companies with fewer than 20 employees were expected to comply by June 2010 [21].

The dietary supplement cGMPs consist of 16 parts and are found in the Code of Federal Regulations at 21 CFR 111, subparts A–P, 111.1 thru 111.610 [26]. They cover all requirements FDA deems necessary to ensure consistent and controlled manufacture of dietary supplements, prevent adulteration or misbranding, and verify the identity, purity, quality, strength, and composition claimed on the product label (Table 1.2). It is important to note that the dietary supplement cGMPs located in 21 CFR Part 111 only apply to facilities that manufacture, package, label, or hold dietary supplements for distribution in the United States. Ingredient suppliers, or those that manufacture only raw ingredients and not finished products, are required to adhere to the food cGMPs located in 21 CFR Part 110; however, some elect to adhere to both cGMP rules as a best practice. In addition, some dietary supplement manufacturers have designed or adhere to other third-party systems that go above and beyond the cGMP requirements. Since dietary supplements are regulated under the "umbrella" of food, manufacturers are required to comply with all other applicable regulations. These include the requirements set forth in the FSMA of 2010 [27] and the Public Health Security and Bioterrorism Preparedness and Response Act of 2002 [28], among others.

Dietary supplements are, by design, a unique hybrid between foods and drugs. Even though dietary supplement cGMPs are modeled after those established for food, many similarities exist between dietary supplement and drug cGMPs, such as identity testing of incoming ingredients (not required for conventional foods). Like food cGMPs, dietary supplement cGMPs address sanitary production practices, appropriate quality control operations, and prevention of adulteration during

Table 1.2 Key elements of the dietary supplement current good manufacturing processes rule

Subpart	Heading	Key elements
A	General Provisions	Who is subject to the rule; definition of terms; provisions and regulations
B	Personnel	Personnel training, qualification, hygiene, cleanliness
C	Physical Plant and Grounds	Maintenance and sanitization of physical plant and ground, pest control, safe and sanitary water supply, trash and sewage disposal, adequate plumbing, ventilation and lighting
D	Equipment and Utensils	Equipment and utensil calibration, inspections and checks; maintenance and cleaning; appropriate design and construction
E	Requirement to Establish a Production and Process Control System	Production and process control system: establish quality control operations; establish specifications for components, packaging, labels, finished products; confirmation that specifications are met through appropriate testing; rejection of components, products, packaging, or labels that do not meet specifications
F	Production and Process Control: Requirements for Quality Control	Quality control activities: laboratory, equipment, controls, components, packaging, labels, master manufacturing records, batch records, material review, and disposition decisions
G	Production and Process Control: Requirements for Components, Packaging, and Labels and for Product That Is Received for Packaging or Labeling as a Dietary Supplement	Receipt of components, packaging, labels, and products for packaging and labeling
H	Production and Process Control: Requirements for a Master Manufacturing Record	Master manufacturing records: preparation of records to specify components, amounts, controls, steps, packaging, and labeling for each dietary supplement product
I	Production and Process Control: Requirements for a Batch Production Record	Batch production record: documentation of date and time each step in the production processed is performed

(continued)

Table 1.2 (continued)

Subpart	Heading	Key elements
J	Production and Process Control: Requirements for Laboratory Operations	Laboratory operations: establish control processes; criteria for appropriate specifications, sampling plans, and testing methods
K	Production and Process Control: Requirements for Manufacturing Operations	Manufacturing operations: design and selection of manufacturing processes to ensure product specifications are consistently met; prevention of contamination
L	Production and Process Control: Requirements for Packaging and Labeling Operations	Packaging and labeling operations: specifications and controls; complete manufacturing history and control of packaged and labeled product through distribution; repackaging and relabeling
M	Holding and Distributing	Holding and distributing operations: appropriate holding conditions of finished products, packaging, labels, in-process materials, and reserve samples; distributing conditions to protect against contamination and deterioration
N	Returned Dietary Supplements	Returned dietary supplements: quarantine returned products until quality control conducts material review and make disposition decisions; disposal, salvaging, and reprocessing
O	Product Complaints	Product complaints: review and investigation of complaints; investigation of possible failures to meet cGMP specifications
P	Records and Recordkeeping	Records must be kept for 1 year past shelf-life date if one is used or 2 years after date of distribution of final batch of product

manufacturing, packaging, storage, and distribution. Similar to drug cGMPs, dietary supplements cGMPs also include provisions related to ensuring the identity, purity, strength, quality, and composition of the finished product. However, unlike drug cGMPs, there is no specific requirement for validation of test methods and processes for dietary supplements because the available science provides only limited officially validated methods. When used, manufacturers must affirm that testing methods are appropriate and scientifically valid.

The cGMPs also require that specifications be set at any stage in the manufacturing process where necessary to ensure the quality of the dietary supplement. While the cGMPs require specifications to be set for the components used in the manufacture of the dietary supplement, labeling and packaging, finished batches, and products received from a supplier, it does not require specifications to be established for dissolution, disintegration, or bioavailability of the dietary supplement. The rule also does not require an expiration date or shelf date on the product label. However, if one of the aforementioned specifications is established, FDA requires manufacturers to possess the supporting data. For example, if a manufacturer chooses to label a dietary supplement with an expiration date, that date must be validated before it is placed on the product label. In addition, batch production records are required for each batch to allow tracing of products throughout the distribution system if needed, for example, when a product complaint must be investigated for a possible cGMP failure.

Internal standard operating procedures (SOPs) are established procedures that manufacturers must follow when carrying out a given operation or in a given situation. SOPs are required to be written and followed for each of part of the dietary supplement cGMP rule in a standardized format. FDA has stressed that SOPs must be in writing in order to be recognized by the agency. A typical SOP provides a header, purpose, scope, definitions, responsibilities, procedure, approvals, revision history, and appendices, which demonstrates to FDA that a manufacturer's system is properly controlled. The dietary supplement cGMPs do not require a facility to establish and follow written procedures for product recalls; however, FDA encourages companies to refer to its guidance document *Guidance for Industry: Product Recalls, Including Removal and Corrections* [31].

Equally important as establishing standardized SOPs is the maintenance of records and a recordkeeping system. Written records must be kept for one year past the shelf-life date (if shelf-life dating is used), or two years beyond the date of distribution of the last batch of dietary supplements associated with those records. Records must be kept as original records, as true copies (e.g., microfilm or photocopies), or as electronic records. Electronic records must comply with 21 CFR Part 11 [32], which defines the criteria under which electronic records and electronic signatures are considered to be trustworthy, reliable, and equivalent to paper records. Written procedures are required for the following operations: cleaning the physical plant and pest control; calibrating instruments, and maintaining and cleaning all equipment and utensils; quality control such as conducting material review and making disposition decisions; receiving components, packaging, and labels; laboratory operations such as testing and examinations to determine whether specifications are met; manufacturing operations, packaging and labeling operations, and holding and distribution operations; and quarantine of returned products and handling customer complaints. These records provide necessary testing results and help demonstrate to FDA the company's compliance with cGMPs. Further, any errors should be corrected by placing a line through the error (where it is still legible), along with the initials of the appropriate individual and date. All records, or copies of records, must be readily available during the retention period for inspection. Facilities must also allow FDA to photocopy records when requested.

FDA has the legal authority to perform an audit in order to ensure companies are conforming to cGMP requirements. Actions that FDA may take upon and after an inspection include the following:

- Form 482—A "notice of inspection" that is presented with investigator credentials upon arriving to the plant
- Form 483—Is issued to firm management at the conclusion of an inspection when an investigator(s) has observed any conditions that in their judgment may constitute violations of the FDCA and related Acts
- Form 484—A "receipt for samples" is presented when an investigator collects samples from the site
- Establishment Inspection Report (EIR)—Is presented within 30 days of an inspection designating one of the following: (1) no action indicated (NAI); (2) voluntary action indicated (VAI); (3), or official action indicated (OAI)
- Warning Letter—Is issued especially in the case of serious findings or if the response to Form 483 is classified as inadequate
- Consent Decree—A legal arrangement with the firm to make specific changes; the agreement is enforceable by federal courts. Consent decrees are only generally sought by FDA after a company has repeatedly violated cGMP requirements
- Mandatory Recall—The FSMA provides FDA with authority to issue a mandatory recall when a company fails to voluntarily recall unsafe food, including dietary supplements, after being asked to by FDA
- Product Seizure
- Fines
- Indictment

These tools allow FDA to ensure the safety of the food supply (inclusive of dietary supplements). In addition, FDA frequently works with industry and state partners to publish press releases and other public notices about recalls that may potentially present a significant or serious risk to the consumer or user of the product. For example, FDA recently published a consumer update on its website warning consumers to avoid use of dietary supplements containing dimethylamylamine (DMAA) [29].

The main subparts of the cGMPs (part A-P) relate to personnel, physical plants, equipment and utensils, production and process controls, holding and distribution, returned dietary supplements, and product complaints. In 2010, FDA published a guidance document with nonbinding recommendations to assist manufacturers with complying with the dietary supplement cGMP rule [30].

1.2.1 Personnel

Manufacturers must ensure that they have qualified employees and supervisors, procedures to remove persons from operation who may be a source of contamination (e.g., an employee presenting an illness), and must implement hygienic practices in place to protect against contamination (e.g., use of gloves, hairnets, outer garments such as gowns or aprons, and hand-washing).

1.2.2 Physical Plants

Physical plants must be clean and sanitary and of appropriate design and construction to prevent dietary ingredients and supplements from becoming adulterated with contaminants during manufacturing, packaging, labeling, and holding operations.

1.2.3 Equipment and Utensils

Equipment and utensils must be appropriate in design and construction for their intended use and must be adequately cleaned, sanitized, and maintained. Instruments and controls must be calibrated for accuracy and precision.

1.2.4 Production and Process Controls

Manufacturers must establish a quality control unit. Quality control personnel responsibilities include approving or rejecting all materials, processes, controls, labels, and finished products and ensuring that testing is conducted on established specifications. The quality control unit must also review and approve all written procedures and specifications, and make disposition decisions regarding components or products that do not meet specifications.

Manufacturers must prepare and follow master manufacturing records to ensure production of uniform products; these records specify the components, quantities, controls and procedures, packaging and labels, and procedures for sampling and testing for each product. Batch records must also be kept for every batch produced to document the date and time of performance of each step in the manufacturing process as well as the results of tests or examinations performed. Incoming shipment of components, dietary ingredients, in-process materials, or dietary supplements must be tested for adherence to specifications. Finished products must be tested based on a statistically sound sampling plan, if testing is not done on every batch.

1.2.5 Holding and Distribution

Manufacturers must establish measures to protect components, dietary ingredients, dietary supplements, packaging, and labels against contamination and deterioration. In addition, they must ensure proper holding conditions including, but not limited to, appropriate temperature, humidity, and light.

1.2.6 Returned Dietary Supplements

This rule requires identification and quarantine of any returned dietary supplement until quality control personnel conduct a material review and make a disposition decision. A returned dietary supplement should be destroyed or otherwise suitably disposed of unless quality control personnel approve the salvage of the returned dietary supplement for redistribution or approve it for reprocessing.

1.2.7 Product Complaints

FDA encourages firms to investigate all complaints in a consistent manner, regardless of whether the complaints relate to the quality of the dietary supplement or to the inherent safety of a dietary ingredient. Manufacturers must also keep written records of product complaints related to cGMPs. A firm is required to investigate a product complaint when there is a reasonable possibility of a connection between consumption of the dietary supplement and an adverse event.

1.3 Safety Standards

1.3.1 Dietary Supplement Safety

As a category of food, the regulations related to the safety of dietary supplements are similar to those for conventional food in several important ways. Namely, dietary supplement manufacturers are not required to register their products with FDA, or obtain FDA approval before producing or selling their products. Prior to marketing a product, however, manufacturers must ensure that a dietary supplement and its ingredients are safe. And like all FDA-regulated products, FDA has authority to take enforcement action if a dietary supplement is found to be unsafe. There are also several distinct requirements for dietary supplements based on history of use, and the recognition that the safety profile and intended use of dietary supplements differs from drugs. These laws include protections for consumers against potentially unsafe products and ingredients. In addition to dietary supplement-specific cGMPs, FDA monitors the safety of dietary supplements through post-market surveillance of mandatory adverse event reports (AERs), which include serious injuries or illnesses that may be related to the product. FDA also monitors the labeling and claims for dietary supplements to make sure the information is truthful and accurate. These additional safety measures, along with the basic legal requirements for both food and supplements, balance FDA's need to protect the food supply with the need for continued consumer access to dietary supplements.

Prior to the passage of DSHEA in 1994, the term *dietary supplement* was not formally defined in law. Thus, FDA regulated vitamins, minerals, botanicals, and other dietary supplement ingredients (and combinations of these ingredients) either as a conventional food—a food additive or GRAS ingredient—or as a drug. DSHEA reaffirmed that dietary supplements should be regulated as a category of food and created a specific definition for dietary supplements. This designation avoided a process whereby dietary supplements would be subject to a lengthy premarket approval process, which is appropriately reserved for products such as drugs and medical devices. Congress, industry, and other interested stakeholders debated this classification and the appropriate framework extensively in the years leading up to DSHEA. However, it was determined that because dietary supplements have been safely consumed by millions of Americans for many years, a drug-like, premarket approval process was unnecessary. Furthermore, the added regulatory requirements would significantly increase the cost of these products. Since the safety profile for these products is well established and vastly different than that of new, novel pharmaceutical ingredients, such a system would not be justified for dietary supplements, particularly from an FDA resources standpoint.

DSHEA also clarified that ingredients of dietary supplements could not be regulated as food additives by specifically excluding dietary supplements from the food additive provisions of the FDCA. A food additive is any substance added to food intended to affect the characteristic of a food, and it requires premarket approval unless the ingredient is GRAS or the ingredient is otherwise approved for use by FDA or the U.S. Department of Agriculture (USDA) [33]. This definition includes any substance used in the production, processing, treatment, packaging, transportation, or storage of food. Unlike dietary supplement ingredients, food additives are added to food for nonnutritive purposes. Dietary supplements and their ingredients are intended to provide general and specific health benefits to consumers by supplementing the diet with additional essential nutrients and other beneficial compounds. Thus, food additive regulations are not suited to address dietary supplement ingredients and the role of dietary supplements in the diet. Congress also held the view that dietary supplements are a low-risk product category, and therefore sought to impose a safety standard for dietary ingredients that reflected these differences.

Congress also recognized the role of dietary supplements as part of a healthy lifestyle and their contribution to preventative health goals. When used safely and appropriately, dietary supplements can help limit the incidence of chronic disease and promote overall good nutrition, thereby reducing long-term healthcare expenditures [2]. Classifying dietary supplements as drugs or otherwise imposing a burdensome, preapproval process would create regulatory obstacles and likely impede the flow of information to consumers about these products, and therefore limit access to an important and affordable means of maintaining health. In promulgating the dietary supplement safety regulations, FDA drew from the various existing paradigms for both food and drug safety, but also heeded Congress's clear message that supplements should still be regulated as a food without unnecessary regulatory hurdles. The current system created as a result of DSHEA provides that the manufacturer (or distributor) and FDA share responsibility for the safety of dietary supplements.

As with any food product, FDA has legal authority to take immediate enforcement action against a dietary supplement that is adulterated (unsafe). A product is considered to be unsafe under the FDCA if it "may be injurious to health," [34] which is applicable to all foods and is the fundamental authority that has permitted FDA to ensure the safety of the food supply since 1906. However, additional adulteration provisions of the FDCA, as amended by DSHEA, provide the primary safety mechanisms for ensuring dietary supplement safety and FDA's enforcement authority. Because FDA does not conduct or require safety or efficacy studies, or approve dietary supplements before they are marketed, manufacturers must ensure the safety by verifying that products do not pose "significant or unreasonable risk of illness or injury under the conditions of use recommended or suggested in the labeling of the product, or if no conditions of use are suggested or recommended in the labeling, under ordinary conditions of use." [35] This includes assurances that ingredients and the product meet FDA's standards for identity, purity, strength, and composition, and the final product is safe when used as directed.

Furthermore, only permissible dietary ingredients may be used in supplements, as DSHEA requires premarket notification to FDA for certain new dietary ingredients (NDIs) [36]. Dietary ingredients already on the market prior to the 1994 passage of DSHEA, known as "old dietary ingredients" (ODIs), were excluded from the definition of an NDI [37]. This premarket notification to FDA must occur at least 75 days before marketing the new ingredient and must include information demonstrating that the ingredient "is reasonably expected to be safe." [38] In contrast, for a food additive to be approved for use, a manufacturer must demonstrate "reasonable certainty of no harm under conditions of intended use." [39] DSHEA also provided FDA with the ability to remove products from the market if the agency finds that a product causes an "imminent hazard" or an immediate safety concern [40]. In addition, Section 403(g)(1) of the FDCA states dietary supplement may also be considered adulterated if it has been prepared, packed, or held under conditions that do not meet cGMP regulations or applicable expiration date labeling regulations. Whenever FDA exercises its enforcement authority under these adulteration provisions, the statute is clear that FDA has the burden of proof to demonstrate that a dietary supplement is unsafe [41].

FSMA provided FDA with additional tools to help the agency maintain the safety of the food supply, particularly imported food products. The Public Health Security and Bioterrorism Preparedness and Response Act of 2002 (the Bioterrorism Act) initially directed FDA to take steps to protect the public from a threatened or actual terrorist attack on the U.S. food supply and other food-related emergencies.

To carry out certain provisions of the Bioterrorism Act, FDA established regulations requiring that food facilities register with FDA. FSMA was enacted on January 4, 2011, and expanded on this authority by amending the FDCA to require that facilities engaged in manufacturing, processing, packing, or holding food—including dietary supplements—for human or animal consumption in the United States to submit additional registration information to FDA. FSMA requires all food facilities, including dietary supplement companies, to renew their facility registrations every other year [42]. The law also granted FDA the authority to suspend the registration of a food facility if the agency finds there is a reasonable probability of causing serious adverse health consequences or death [43] and access records during a facility inspection [42]. FSMA also provided the agency with mandatory recall authority when a company fails to voluntarily recall an unsafe food after being requested to do so by FDA [44], and provides the agency with the ability to detain potentially unsafe products [45].

Finally, FDA also monitors the labeling of dietary supplements to ensure that they are truthful and not misleading, and that they otherwise meet FDA requirements for labeling. This includes product claims as well as identity information about a dietary supplement (e.g., ingredients, serving size). Products that do not meet these requirements are considered misbranded [46] and subject to the same enforcement provisions as adulterated products, including injunction, seizure, recall, fines, and/or civil penalties.

1.3.2 Dietary Ingredient Safety

A primary goal of DSHEA was to create a regulatory framework for dietary supplements that would provide continued access to products that clearly had a history of safe use. The legislation sought to avoid regulatory obstacles and encourage innovation, while balancing the need to ensure the safety of dietary supplement ingredients, especially new or novel ingredients.

As discussed previously, DSHEA classified all dietary ingredients marketed in the United States before October 15, 1994 as ODIs. These "grandfathered" ingredients are considered safe for continued consumer use. Therefore, dietary supplement manufacturers that use ODIs are not required to submit additional safety information to FDA before putting products into the market that contain ODIs. However, the law still requires that manufacturers and distributors ensure that all dietary ingredients, whether new or old, and all dietary supplements do not pose significant or unreasonable risk of injury.

In contrast, a dietary supplement containing a NDI is considered adulterated unless (1) "the dietary supplement contains only dietary ingredients which have been present in the food supply as an article used for food in a form in which the food has not been chemically altered" or (2) the manufacturer submits a NDI notification (NDIN) at least 75 days before the product is introduced into the market [47]. The NDIN should include information that establishes that the dietary ingredient "will reasonably be expected to be safe" under the conditions of use recommended in the labeling [38], a standard that is distinctly different from the standards applicable to food additives and conventional food ingredients. The NDIN process involves gathering ingredient scientific evidence that demonstrates a history of safe use of the ingredient and substantially similar ingredients, along with dosage and chemical identity of the NDI, and the recommended use. However, DSHEA did not specify what causes an ingredient to be "chemically altered" and what type of evidence is sufficient to demonstrate that the NDI is safe.

To provide further clarification on the NDIs, FDA released its *Draft Guidance for Industry: Dietary Supplements: New Dietary Ingredient Notifications and Related Issues* (Draft NDI Guidance) in July 2011 [48]. It is intended to assist industry in deciding when a NDIN is necessary and what qualifies as a NDI, the procedures for submitting a notification, the types of data and information that FDA recommends manufacturers and distributors consider when they evaluate the safety of a dietary supplement containing a NDI, and what should be included in the notification. However, following the release of the Draft NDI Guidance, it became clear that the dietary supplement industry and FDA had very different opinions regarding the Congressional intent of DSHEA's NDI provisions. A primary reason for this dissonance was that the Draft NDI Guidance was published nearly 20 years after the passage of DSHEA, making it difficult for industry to prove which ingredients are in fact ODIs and therefore do not require a NDIN. Other areas of the Draft NDI Guidance describe complex areas such as chemical alteration and synthetic botanicals, where some segments of the industry and FDA clash regarding the safety impact of modern techniques on dietary ingredients. To be sure, FDA received over 146,000 pages of comments regarding the Draft NDI Guidance, which came from approximately 7,000 entities and individuals [49]. Furthermore, the main proponents and authors of DSHEA, Senators Orrin Hatch (R-Utah) and Tom Harkin (D-Iowa), voiced their opposition to the guidance and called for its withdrawal [50]. They share industry's concerns that the Draft NDI Guidance represents a swift departure from the goals of DSHEA: to provide consumer with access to safe, affordable products by avoiding a burdensome and unnecessary premarket approval system. Both industry and members of Congress continue to call on FDA to provide a revised guidance that clarifies FDA's intentions and reflects the goals and intent of DSHEA.

The multitude of comments and debate over the key provisions of the guidance make it likely that a final revised draft will take several years. Still, as with other FDA guidance documents, the guidance reflect the agency's "current thinking" and interpretation of the law; it does not have the effect of a law or regulation and neither FDA nor industry must adhere to the document. The Draft NDI Guidance represents recommendations offered by FDA to evaluate the safety of a NDI, but manufacturers can use other mechanisms to satisfy the safety requirements under the law, including the requirements under FSMA and the adulteration provisions described previously.

In addition to verifying that an ingredient is a legitimate ODI or NDI, manufacturers must be aware of other legal requirements that apply to both food and dietary supplements. The FDCA's general adulterations prohibit: "any poisonous or deleterious"; "any filthy, putrid, or decomposed substance"; products that have been "prepared, packed, or held under insanitary conditions"; and the addition of unapproved food additives [51]. The law also prohibits the introduction into commerce of any food that contains an approved drug or a licensed biologic [52] and the use of unapproved color additives [53]. If FDA determines that a product contains a prohibited ingredient or is otherwise unsafe, the agency has the authority to declare that an ingredient, or a dietary supplement, presents an imminent hazard to public health or safety [40]. The marketer or manufacturer/distributor is then required to cease marketing of the product. As noted above, FSMA also provides FDA with the authority to issue a mandatory recall in the case of an unsafe product, or restrict the movement of unsafe food product by suspending a facility's registration or through administrative detention while the product is in transit [44, 45].

1.3.3 Adverse Event Reporting

Dietary supplement safety is also monitored through adverse event reporting and related recordkeeping requirements. Adverse events are health-related events, such as an allergic reaction, that occur following the use of product, and may or may not be associated with the product. Reporting of adverse events is required for certain FDA-regulated products, including drugs, vaccines and other biologics, medical devices, and most recently, dietary supplements and over-the-counter medicines.

As a category of food, the regulation of dietary supplements did not initially include mandatory adverse event reporting. With the support of the dietary supplement industry, Congress enacted the Dietary Supplement and Nonprescription Drug Consumer Protection Act on December 22, 2006 [54]. The law requires manufacturers, packers, and distributors of dietary supplements in the United States to report to FDA all serious adverse effects associated with the use of these supplements. The law defines serious adverse events as an adverse event that "results in death; a life-threatening experience; inpatient hospitalization; a persistent or significant disability or incapacity; or a congenital anomaly or birth defect; or requires, based on reasonable medical judgment, a medical or surgical intervention to prevent an outcome described." [55] Each dietary supplement label must include the contact information (the domestic address or phone number) for the supplement's manufacturer, packer, or distributor—referred to as the "responsible person" who is charged with reporting serious adverse events to FDA [56]. The responsible person must submit a serious adverse event to FDA no later than 15 business days after the report is received; any new medical information related to a previously submitted adverse event report must be submitted to the agency within one year of the initial report, and within 15 days of receipt [57]. The law also requires companies to maintain records related to all AERs, both serious and non-serious, for a period of six years, and authorizes FDA to inspect these records to ensure compliance with these recordkeeping requirements [58].

FDA has been clear that AERs do not demonstrate causality between a product and the adverse event. Rather, the agency uses adverse events as early warning signals that suggest potential safety concerns with a product, such as contamination, adulteration, tampering, or ingredient safety issues, which can be followed up with a more detailed investigation. FDA's Center for Food Safety and Applied Nutrition (CFSAN) Adverse Event Reporting System (CAERS) collects AERs and product complaints for dietary supplements, as well as voluntary reports for conventional foods and cosmetics. However, the requirement to report serious adverse events applies only to dietary supplements. Qualified CFSAN staff review and analyze these adverse event submissions to determine if further agency action is necessary. This type of post-market surveillance helps confirm the safety of a product, as many FDA-regulated products are initially approved or tested based on limited clinical trials that are not representative of the general population. Monitoring of adverse events can also help refine the safety profile of a product after it is introduced into the market and used by a larger, more diverse population. In the case of dietary supplements, if CAERS determines there is a causal relationship between the product and the adverse event, FDA will take appropriate steps to mitigate further risk to consumers.

1.4 Labeling

1.4.1 Addressing Labeling and Claims

Like conventional food and other FDA-regulated products, dietary supplements must be properly labeled to be legally marketed in the United States. DSHEA amended the FDCA by adding specific labeling requirements for dietary supplements. Three years later, FDA finalized regulations that address the statement of identity, nutrition labeling, ingredient labeling, and nutrient content and health claims for dietary supplements [22]. The obvious difference between the labeling of dietary supplements and conventional foods is that supplements include a Supplement Facts panel in lieu of the Nutrition Facts panel used for conventional foods. The Supplement Facts panel (also called the Supplement Fact box) and Nutrition Facts panel are similar in format and layout for ease of consumer understanding, but FDA imposes distinct requirements for the two types of labels. Likewise, the claims that can be made for food and dietary supplement are also similar. However, marketers must be aware of the unique requirements and limitations on the claims that can be made for each of these product categories.

FDA is the agency with primary responsibility for the regulation of dietary supplement labels, ensuring that required label components are listed correctly and that the label accurately describes the product's ingredients. The product label itself identifies the product as a dietary supplement, although FDA refers to "labeling" in a broader sense to include any marketing materials that accompany the product.

Thus, FDA oversight of product labeling extends beyond the physical product label and includes all forms of labeling, such as marketing, packaging material, inserts, and other promotional materials distributed at the point of sale. Although Section 5 of DSHEA specifically excludes certain third-party information from the definition of labeling, including articles, abstracts, or book chapters, this information must be truthful, not misleading, and published by an independent entity other than the manufacturer or marketer promoting the product [2].

FDA and the Federal Trade Commission (FTC) share responsibility for the regulation of dietary supplement advertising claims pursuant to a long-standing liaison agreement [59]. FDA focuses on claims included on the product label itself and any accompanying promotional or informational material, as well as the content of the claims to ensure the language complies with regulatory requirements. FTC focuses primarily on advertising claims and ensuring that claims are truthful and not misleading to consumers and have adequate scientific support. Both agencies monitor claims made through Internet advertising and will often coordinate enforcement priorities in this area.

Because dietary supplements are a category of food, the general labeling requirements for these products are similar; all product labels must be accurate, easy to read, and specific to the product. Section 403(s) of the FDCA [60], as amended by DSHEA, sets forth the basic labeling requirements for dietary supplements. The FDA regulations implementing this statute prescribe the more detailed requirements, such as font, print size, and placement of the label. All dietary supplement labels must include a statement of identity; the net quantity of contents; nutrition labeling, including the Supplement Facts panel; a complete list of product ingredients; and the name and place of business of the manufacturer, packer, or distributor. If a dietary supplement label fails to include all required information in the correct format, FDA may deem the product misbranded under the FDCA [46]. FDA may also require additional label information and/or disclaimer statements for dietary supplements that meet specific criteria.

1.4.1.1 Statement of Identity and Net Quantity of Contents

The statement of identity is the common name of the product, or a term that adequately describes the product. It must include the word "dietary supplement," although the word "dietary" may be substituted by the name of the dietary ingredient(s) in the product. Examples include "vitamin D supplement," "calcium supplement," and "herbal supplement with vitamins." FDA requires that the statement of identity be displayed as one of the most prominent features on the label. The "net quantity of contents" statement informs consumers how much of the dietary supplement is contained in the package and must be expressed, either in weight or measure, numerical count, or a combination thereof (e.g., "60 capsules" or "90 soft gels—1,000 mg"). If the quantity is provided in weight or measure, it must be expressed using both metric units and the U.S. customary system (e.g., "net wt. 18 oz. (510 g)").

The statement of identity and net quantity of contents statement must appear on the front label panel, which is referred to as the principal display panel. This is the portion of the package facing outward on a store shelf and most likely to be seen by the consumer. If the package has two or more appropriate surfaces, the statement of identity and net quantity may be placed on alternate principal display panels. FDA regulations provide further detail regarding how to locate the principal display panel and the placement of the statement of identity and net quantity of contents, and required type size. [61]

1.4.1.2 Nutrition Labeling

The information panel is the label panel immediately to the right to the principle display panel and provides consumers with information about the ingredients contained in a dietary supplement, ingredient amounts, nutrition information, and serving size. The information panel must include a prominent box with the heading "Supplement Facts," similar to the Nutrition Facts panel found on conventional food products. FDA regulations provide special labeling provisions for small or intermediate size packaging and exemptions from the nutrition labeling requirements for firms that sell small quantities of supplements or that ship bulk quantities of products, or are otherwise exempt [62].

The Supplement Facts box identifies each dietary ingredient in the product and the quantity of each ingredient. Unlike conventional foods, the source of the dietary ingredient may be listed in more than one location. For example, the amount of the dietary ingredient eicosapentaenoic acid (EPA) may be listed in the Supplements Facts box (e.g., EPA, 330 mg). However, the source of the EPA (purified fish oil) may be listed either the Supplements Facts box *or* alternatively in the ingredients statement that is located outside of the Supplement Facts box (see Ingredients Statement below). One exception to this regulation is that for botanical ingredients, the part of the plant from which the dietary ingredient is derived must be listed in the Supplement Facts box. FDA regulations also provide specific details regarding formatting requirements, such as the required type size [63].

Immediately below the Supplement Facts heading (see sample panels in Fig. 1.1), the panel must list the "serving size" and "servings per container" (e.g., "Serving Size: 3 Tablets"). The serving size of a dietary supplement is the maximum amount recommended on the label for consumption per eating occasion; if no recommendations for usage are provided, the serving size is equal to one unit. If the serving size is provided as a range (e.g., "Serving Size: 1–3 tablets." or "Serving Size: 2–3 tablets 3 times daily."), the serving size would be the maximum amount specified in the provided range. In these examples, the serving sizes would be 3 tablets and 9 tablets, respectively. Although manufacturers may choose how to express serving size, the information must be listed on the Supplement Facts box in a manner that is consistent with the product's directions for use and the product formulation. The phrase "servings per container" is not necessary if the net quantity of contents statement provides identical information. For example, if the net quantity of contents is listed

a

Supplement Facts

Serving Size 1 Tablet

	Amount Per Serving	% Daily Value
Vitamin A (as retinyl acetate and 50% as beta-carotene)	5000 IU	100%
Vitamin C (as ascorbic acid)	60 mg	100%
Vitamin D (as cholecalciferol)	400 IU	100%
Vitamin E (as dl-alpha tocopheryl acetate)	30 IU	100%
Thiamin (as thiamin mononitrate)	1.5 mg	100%
Riboflavin	1.7 mg	100%
Niacin (as niacinamide)	20 mg	100%
Vitamin B_6 (as pyridoxine hydrochloride)	2.0 mg	100%
Folate (as folic acid)	400 mcg	100%
Vitamin B_{12} (as cyanocobalamin)	6 mcg	100%
Biotin	30 mcg	10%
Pantothenic Acid (as calcium pantothenate)	10 mg	100%

Other ingredients: Gelatin, lactose, magnesium stearate, microcrystalline cellulose, FD&C Yellow No. 6, propylene glycol, propylparaben, and sodium benzoate.

b

Supplement Facts

Serving Size 1 Tablet

Amount Per Serving		% Daily Value for Children Under 4 Years of Age	% Daily Value for Adults and Children 4 or more Years of Age
Calories	5		
Total Carbohydrate	1 g	†	< 1%*
Sugars	1 g	†	†
Vitamin A (50% as beta-carotene)	2500 IU	100%	50%
Vitamin C	40 mg	100%	67%
Vitamin D	400 IU	100%	100%
Vitamin E	15 IU	150%	50%
Thiamin	1.1 mg	157%	73%
Riboflavin	1.2 mg	150%	71%
Niacin	14 mg	156%	70%
Vitamin B_6	1.1 mg	157%	55%
Folate	300 mcg	150%	75%
Vitamin B_{12}	5 mcg	167%	83%

* Percent Daily Values are based on a 2,000 calorie diet.
† Daily Value not established.

Other ingredients: Sucrose, sodium ascorbate, stearic acid, gelatin, maltodextrins, artificial flavors, dl-alpha tocopheryl acetate, niacinamide, magnesium stearate, Yellow 6, artificial colors, stearic acid, palmitic acid, pyridoxine hydrochloride, thiamin mononitrate, vitamin A acetate, beta-carotene, folic acid, cholecalciferol, and cyanocobalamin.

c

Supplement Facts

Serving Size 1 Packet
Servings Per Container 10

	AM Packet		PM Packet	
Amount Per Serving		% Daily Value		% Daily Value
Vitamin A	2500 IU	50%	2500 IU	50%
Vitamin C	60 mg	100%	60 mg	100%
Vitamin D	400 IU	100%		
Vitamin E	30 IU	100%		
Thiamin	1.5 mg	100%	1.5 mg	100%
Riboflavin	1.7 mg	100%	1.7 mg	100%
Niacin	20 mg	100%	20 mg	100%
Vitamin B_6	2.0 mg	100%	2.0 mg	100%
Folic Acid	200 mcg	50%	200 mcg	50%
Vitamin B_{12}	3 mcg	50%	3 mcg	50%
Biotin			30 mcg	10%
Pantothenic Acid	5 mg	50%	5 mg	50%

Ingredients: Sodium ascorbate, ascorbic acid, calcium pantothenate, niacinamide, dl-alpha tocopheryl acetate, microcrystalline cellulose, artificial flavors, dextrin, starch, mono- and diglycerides, vitamin A acetate, magnesium stearate, gelatin, FD&C Blue #1, FD&C Red #3, artificial colors, thiamin mononitrate, pyridoxine hydrochloride, citric acid, lactose, sorbic acid, tricalcium phosphate, sodium benzoate, sodium caseinate, methylparaben, potassium sorbate, BHA, BHT, ergocalciferol and cyanocobalamin.

d

Supplement Facts

Serving Size 1 Capsule

Amount Per Capsule	% Daily Value
Calories 20	
Calories from Fat 20	
Total Fat 2 g	3%*
Saturated Fat 0.5 g	3%*
Polyunsaturated Fat 1 g	†
Monounsaturated Fat 0.5 g	†
Vitamin A 4250 IU	85%
Vitamin D 425 IU	106%
Omega-3 fatty acids 0.5 g	†

* Percent Daily Values are based on a 2,000 calorie diet.
† Daily Value not established.

Ingredients: Cod liver oil, gelatin, water, and glycerin.

Fig. 1.1 Sample dietary supplement labels. (**a**) Dietary supplement containing multiple vitamins (see 21 CFR 101.36(e)(10)(i)). (**b**) Dietary supplement containing multiple vitamins for children and adults (see 21 CFR 101.36(e)(10)(ii)). (**c**) Multiple vitamins in packets (see 21 CFR 101.36(e)(10)(iii)). (**d**) Dietary supplement containing dietary ingredients with and without RDIs and DRVs (see 21 CFR 101.36(e)(10)(iv)

as 60 tablets and the serving size is 1 tablet, the servings per container is 60; this redundant information may be omitted from the Supplement Facts box.

Inside the Supplement Facts box and immediately following the serving size information (see sample labels in Fig. 1.1), dietary ingredients that have an established Reference Daily Intake (RDI) or Daily Reference Value (DRV) must be listed when present in measurable amounts. FDA states that a "measurable amount" is that which exceeds the amount that can be declared as zero in nutrition labeling of foods,

Table 1.3 Dietary ingredients that contribute amounts in excess of the following listed amounts must be declared on the Supplement Facts panel

Category	Amount
Total calories	5 kcal
Calories from fat	5 kcal
Total fat	0.5 g
Saturated fat	0.5 g
Cholesterol	2 mg
Sodium	5 mg
Total carbohydrate	0.5 g
Dietary fiber	0.5 g
Sugars	0.5 g
Protein	0.5 g
Vitamin A	2 % DV
Vitamin C	2 % DV
Calcium	2 % DV
Iron	2 % DV

in accordance with 21 CFR §101.9(c). Dietary ingredients that contribute amounts in excess of the listed amounts (based on one serving, as indicated in Table 1.3) must be declared on the product label. If the protein content is composed only of individual amino acids, these must be listed separately and not as protein.

Vitamins and minerals that are considered essential to human nutrition, other than those listed in Table 1.3, such as vitamin D, magnesium, and folate, must only be declared when intentionally added for purposes of supplementation (above 2 % RDI or DV). These nutrients must also be declared if a claim is made about them. FDA regulation 21 CFR §101.9(c)(8)(iv) and (c)(9) provides the full list of additional vitamins and minerals. Additional FDA regulations also describe the order in which these nutrients must appear and the terminology that must be used on the label [64]. Furthermore, if the product makes a claim about calories from saturated fat, insoluble fiber, polyunsaturated fat, sugar alcohol, monounsaturated fat, other carbohydrate, or soluble fibers, these nutrients must also be listed in the Supplement Facts box (see sample labels in Fig. 1.1).

Dietary ingredients listed in the Supplement Facts box must also provide the amount per serving, which must be consistent with the statement of serving size (e.g., "Amount per serving," "Amount per 3 capsules," or "Amount per teaspoon") (see sample labels in Fig. 1.1). If the label recommends multiple servings daily, the amounts listed on the label may reflect information for more than one serving (e.g., if the label provides instructions for A.M. and P.M. dosage amounts).

The Supplement Facts box must list the % DV of all dietary ingredients, with the exception of protein or dietary ingredients that do not have an established DRV (e.g., sugar), or if the label states that the product is intended for infants, children younger than 4 years of age, or pregnant or lactating women. Dietary ingredients without an established DV must be listed below those with a DV, and must include the quantity and a symbol in the "% DV" column with a footnote that indicates "Daily Value Not Established" (see sample labels in Fig. 1.1). FDA regulations describe in detail how these percentages must be calculated and listed on the label [65, 66].

Proprietary blends must be identified under the heading "Proprietary Blend" or a similar term. Ingredients with an RDI or DRV must be listed separately from those without an RDI or DRV. Non-RDI or DRV ingredients should be listed in descending order by weight, followed by a symbol indicating "Daily Value Not Established." In order to protect any proprietary interest in the blend, only the total weight of the blend must be declared, rather than the weight of each individual ingredient.

For liquid extracts, the label must list the volume or weight of the total extract and the condition of the starting material prior to extraction. Information regarding the concentration of the ingredient or solvents used is voluntary, although the name of the solvent must be identified in the either the Supplement Facts panel or the ingredient list. For dried extracts (dietary ingredients from which the solvent has been removed), the weight of the dried extract should be listed.

1.4.1.3 Ingredients Statement

If a dietary supplement includes ingredients not listed on the Supplement Facts box, these ingredients must be listed under the heading "Other Ingredients," in accordance with §101.4(g). These ingredients may include the source of dietary ingredients that have not been identified in the Supplement Facts box (e.g., rose hips as the source of vitamin C), other food ingredients (e.g., water and sugar), and technical additives or processing aids (e.g., gelatin, excipients, starch, colors, stabilizers, preservatives, and flavors).

Unlike labeling for conventional foods, if all of the ingredients that are sources of the dietary ingredients are identified in the Supplement Facts box, and there are no other ingredients such as excipients or fillers, an ingredients statement is not necessary. For example, if calcium appears in the Supplement Facts box as "Calcium (as calcium carbonate)," this ingredient may be omitted from the ingredients statement. If the source of the dietary ingredient is not listed in the Supplement Facts box, it must be listed under the heading "Ingredients," in accordance with §101.4(g). For ingredients with multiple sources, all of the sources should be listed within parentheses in descending order by weight. Source ingredients should be identified using their common or usual name, and the listing of botanicals should specify the part of the plant from which the ingredient is derived.

1.4.1.4 Name and Place of Business

The Dietary Supplement and Nonprescription Drug Consumer Act, signed into law on December 22, 2006, requires that dietary supplement labels include contact information (the domestic address or phone number) for the "responsible person" who can receive reports of adverse events associated with the dietary supplement[1] [67].

[1] The statute requires that manufacturers maintain reports of all adverse events, but only serious adverse must be referred to FDA. This requirement is discussed in further detail in Sect. 1.3.

The "responsible person" is typically the manufacturer, packer, distributor, or retailer identified on the dietary supplement label. The statute further requires that this contact information be complete and sufficient to ensure that consumers can readily contact the responsible person if an adverse event occurs. Therefore, if a domestic address is listed on the label, it must include the street address or P.O. Box, city, state, and zip code of the responsible person, and the domestic phone number must include the toll-free or local area code. FDA may deem a dietary supplement misbranded under the FDCA if the labels fails to include the required information, or if the information is incomplete. FDA guidance also recommends, but does not require, that manufacturers also include a clear, prominent statement informing consumers that they may use this contact information to report adverse events [68].

1.4.1.5 Additional Mandatory Label Statements

FDA, and at least one state, requires additional information on the labels of dietary supplements containing certain ingredients. In 1997, FDA issued a final rule requiring that dietary supplements with added iron bear the following label warning statement:

> WARNING: Accidental overdose of iron-containing products is a leading cause of fatal poisoning in children under 6. Keep this product out of reach of children. In case of accidental overdose, call a doctor or poison control center immediately.

The statement must be placed in box and displayed prominently on the product's information panel and any outer packaging. However, this requirement applies only to iron-containing supplements in solid oral dosage form, such as capsules or tablets [69]. In 2003, FDA published a regulation mandating that conventional foods and dietary supplements declare the amount of *trans* fat in the product when present at 0.5 g or more. This information must be listed in the Supplement Facts panel of the dietary supplement under the listing of saturated fat [70, 71].

California's Proposition 65, the Safe Drinking Water and Toxic Enforcement Act of 1986, is a ballot initiative intended to protect California citizens from chemicals known to cause cancer, birth defects, or other reproductive harm, and to inform citizens about exposures to such chemicals. Proposition 65 requires the Governor of California to publish, at least annually, a list of chemicals known to the state to cause cancer or reproductive toxicity. Consumer product labels, including those for dietary supplements, must include the following statement if the product contains one of these listed chemicals in amounts that exceed defined safety limits:

> WARNING: This product contains a chemical known to the State of California to cause cancer [or birth defects or other reproductive harm.]

Companies may be subject to enforcement action by the State Attorney General or civil lawsuits brought by private citizens for failing to provide the required warning.

1.4.1.6 Disclaimers and Disclosures

As required by DSHEA, dietary supplement labels that bear a structure/function claim or nutrient content claim must include the following disclaimer or disclosure statement adjacent to the claim and printed in bold:

> **This statement[s] has not been evaluated by the Food and Drug Administration. This product is not intended to diagnose, treat, cure or prevent any disease.**

FDA does not require this disclaimer for conventional food labels with structure/function claims. Structure/function claims or nutrient content claims are described in further detail in the next section. FDA's *Guidance for Industry: A Food Labeling Guide* provides further detailed information on how to comply with labeling and disclosure requirements for products that make claims [72].

To ensure that nutrients associated with unwanted health outcomes are prominently disclosed, FDA requires dietary supplements that make nutrient content claims (e.g., "high in fiber" or "low in fat") provide a disclosure statement when a nutrient, other than the nutrient described in the claim, exceeds certain levels. The statement is intended to inform consumers that one or more nutrients in the product that may increase the risk of a diet-related disease or health condition (e.g., "See nutrition information for sodium content."). This disclosure statement is required when a product makes a nutrient content claim and the product also contains one or more of the following nutrients in excess of the listed levels, per labeled serving (or per 50 g per serving): fat, 13.0 g; saturated fat, 4.0 g; cholesterol, 60 mg; or sodium, 480 mg. FDA regulation 21 CFR 101.13(h)(1)-(3) provides additional details and requirements for these disclosures.

FDA also permits companies to include additional product information on the label, such as quality assurance statements related to GMP or third-party certification. As with all label claims, these statements must be truthful, accurate, and not misleading to the consumer.

1.4.2 Dietary Supplement Claims

Claims are an important aspect of dietary supplement labeling because they communicate to consumers how the product is intended to be used. By definition, dietary supplements are intended to supplement the diet by providing an additional source of nutrients. Intended use is a key element when FDA determines how the product should be regulated—as a dietary supplement, conventional food, or drug. Even if a product is labeled as a dietary supplement and otherwise meets the statutory definition, FDA may still consider it misbranded or an unapproved drug if the product claims to treat, prevent, cure, or mitigate a disease, which are claims only permitted for drugs. To legally market a product as a dietary supplement and avoid potential enforcement action, any product claims—whether express or implied—must be consistent with DSHEA and FDCA.

Companies must also ensure that claims are adequately substantiated and consistent with both FDA and FTC requirements, which outline the level and quality of scientific evidence required for substantiation. As described previously, advertising claims used to market and promote a dietary supplement fall mainly under the jurisdiction of FTC, but both agencies have enforcement authority over such claims. The dual regulatory oversight of dietary supplement claims by both FDA and FTC may be confusing to dietary supplement marketers. Where appropriate, FDA and FTC have tried to align regulatory requirements to avoid inconsistent policies. These agencies also provide a number of guidance documents to assist companies with this area of regulation [73, 74].

Dietary supplement manufacturers are permitted by law to make four types of claims for supplements: (1) health claims, (2) qualified health claims, (3) structure/function claims, and (4) nutrient content claims. These claims can be expressed as a written statement, third-party reference, symbol, or vignette. FDA's *Guidance for Industry: A Dietary Supplement Labeling Guide* provides a comprehensive overview of these claims and the requirements for use [75]. Understanding the regulation of dietary supplement claims is crucial for dietary supplement companies because failing to adhere to FDA regulatory requirements may cause the product to be misbranded. The legal consequences of using an unauthorized claim, or using a claim that lacks substantiation may include injunction, product recall, and/or seizure of the product by FDA. FTC also has strong enforcement provisions for violations of its statutory and regulatory requirements, including significant fines, penalties, and consumer redress. FTC may also prohibit a company from future advertising in an entire product category through a consent decree. FTC enforcement is discussed in greater detail later in this chapter.

1.4.2.1 Health Claims

Health claims describe the relationship between a certain dietary ingredient and the reduction of risk of a disease or health-related condition and may be used for both conventional food and dietary supplements. However, only health claims specifically authorized by FDA may be used on product labeling, and the agency provides three ways to submit a claim for authorization. The 1990 NLEA gave FDA authority to allow health claims for food or supplements. NLEA health claims may be used to characterize the relationship between a dietary ingredient or supplement and risk of disease (e.g., "Calcium may reduce the risk of osteoporosis." or "Diets adequate in folate reduce the risk of neural tube birth defects."). A company must first submit a petition to FDA requesting permission to use the health claim, in accordance with 21 CFR §101.70. FDA will evaluate the petition and consider whether the evidence supporting the nutrient/disease relationship meets the agency's significant scientific agreement (SSA) standard. FDA will authorize the health claim if the agency "determines based on the totality of the publicly available evidence (including evidence from well-designed studies conducted in a manner which is consistent with generally recognized scientific procedures and principles), that there is significant scientific agreement among experts qualified by scientific training and experience to evaluate such claims, that the claim is supported by such evidence." [76]

FDA applies this same standard to conventional food health claims, which FDA adopted by regulation for dietary supplement health claims. FDA provides a list of authorized health claims on its website, which also includes the specific parameters for using the health claim as well as model language [77]. However, the high evidentiary standard required to meet FDA's SSA threshold has resulted in few FDA-authorized health claims for dietary supplements. Currently, FDA has authorized only four health claims for dietary supplements.

In 1999, a dietary supplement manufacturer challenged FDA's denial of several dietary supplement health claims after failing to meet the agency's SSA standard. In *Pearson v. Shalala*, the court held that the First Amendment does not permit FDA to reject health claims that the agency determines to be potentially misleading, unless the agency also reasonably determines that no disclaimer would eliminate the potential deception [78]. In September 2003, FDA began considering "qualified health claims" under interim procedures established by the agency following this case [79]. FDA eventually published its final guidance for qualified health claims in 2006 [80]. Three years later, FDA published its *Guidance for Industry: Evidence-Based Review System for the Scientific Evaluation of Health Claims*, which explains the agency's process for evaluating the scientific evidence for both health claims and qualified health claims [81].

Qualified health claims may be used when there is emerging evidence for a relationship between a dietary supplement and reduced risk of disease. Because these claims have less evidentiary support than FDA-authorized health claims, FDA requires qualifying language to indicate this reduced level of evidence. Typically, FDA will require an agency-approved disclaimer that describes the level of scientific evidence used to support the claim. Instead of issuing a regulation that authorizes the use of a certain qualified health claim, FDA will exercise enforcement discretion if the claim meets all general requirements of 21 CFR §101.14. Following the submission of a petition to use a qualified health claim, FDA will then issue a letter to the petitioner indicating that the agency will not object to use of the claim if the claim meets these criteria. In addition to the petitioner, other companies may use the qualified health claim so long as the stated criteria are met. FDA's website provides a current list of qualified health claims and factors considered by the agency when exercising its enforcement discretion [82].

The Food and Drug Administration Modernization Act (FDAMA) of 1997 provides another way for FDA to authorize health claims, whereby health claims may be based on an "authoritative statement" from a scientific body of the U.S. Government or the National Academy of Sciences. Because FDAMA did not specifically provide that dietary supplements may use such statements, FDA has stated that authoritative statements may not serve as the basis for a dietary supplement health claims at this time [83].

1.4.2.2 Structure/Function Claims

Section403(r)(6) of the FDCA, as amended by DSHEA [84], permits claims to be made for dietary supplements if:

> …[T]he statement claims a benefit related to a classical nutrient deficiency disease and discloses the prevalence of such disease in the United States, describes the role of a nutrient or dietary ingredient intended to affect the structure or function in humans, characterizes the documented mechanism by which a nutrient or dietary ingredient acts to maintain such structure or function, or describes general well-being from consumption of a nutrient or dietary ingredient.

Structure/function claims describe the effect of a dietary supplement on the structure or function of the body, the biological mechanism by which a product acts, a benefit related to a nutrient deficient disease, or general well-being from the consumption of a nutrient or ingredient. Examples of structure/function claims include the following: "Calcium builds strong bones." and "Fiber maintains bowel regularity." They are the most widely used and most common type of dietary supplement claim. However, the appropriate use of structure/function is often misunderstood due to their resemblance to health claims and disease claims, coupled with the lack of FDA authorization before use.

Structure/function claims differ from health claims because (1) health claims require FDA evaluation and authorization prior to use, whereas structure/function claims do not require FDA preapproval and (2) structure/function claims may not explicitly or implicitly link the relationship to a disease or health-related condition. In contrast to health claims, these claims describe the role of the substance in maintaining or affecting the normal structure or function of the body. For example, the dietary supplement claim "helps maintain normal blood sugar levels in healthy individuals" would be permissible, because it does not link to a specific disease state but rather to the effect of the supplement on maintaining blood sugar levels already within a normal range.

Marketers of dietary supplements must have credible scientific evidence to substantiate a structure/function claim, which demonstrates that the claim is truthful and not misleading. Marketers must also submit notification to FDA of its structure function claims no later than 30 days after marketing a supplement. When a structure function claim is used, the claim must be accompanied by a statement on the product label indicating that the claim "has not been evaluated by the Food and Drug Administration" and that it is "not intended to diagnose, treat, cure, or prevent any disease." In response to 30-day structure/function claim notifications submitted by marketers, FDA may issue a "Courtesy Letter" if the agency believes that the claim is not in compliance with the parameters set forth by DSHEA and FDCA.

FDA has published detailed guidance documents to assist manufacturers to better understand the somewhat vague differences between permitted structure/function claims, FDA-authorized health claims, and claims that may be used only for FDA-approved drugs. FDA's *Structure/Function Claims, Small Entity Compliance Guide* contains detailed information about the requirements for structure/function claims and provides criteria to assist in distinguishing between these types of claims [85]. FDA regulation 21 CFR 101.93(g) defines disease as "damage to an organ, part, structure, or system of the body such that it does not function properly (e.g., cardiovascular disease), or a state of health leading to such dysfunction (e.g., hypertension); except that diseases resulting from essential nutrient deficiencies (e.g., scurvy, pellagra) are not included in this definition." FDA's final regulation on structure/function claims provides further guidance on avoiding disease claims [86]. FDA also published its *Guidance for Industry: Substantiation for Dietary Supplement Claims Made Under Section 403(r)(6) of the Federal Food, Drug, and Cosmetic Act*, which describes the amount, type, and quality of evidence the agency recommends to substantiate a structure/function claim [87], and is similar to FTC's standard for evaluating substantiation for advertising claims.

1.4.2.3 Nutrient Content Claims

Nutrient content claims characterize the level of a nutrient or dietary substance in the product, either expressly or by implication. Examples include "high in antioxidants," "low calorie," "more fiber," and "reduced sugar." However, most nutrient content claims are approved for conventional food products because the regulations pertaining to these claims apply to nutrients with an established DV. Dietary supplements are permitted to make claims regarding the percentage level of dietary ingredients without an RDI or DRV, but these claims cannot use terms such "high" or "low." In addition, any comparative references to other products must be specific and clearly defined. Dietary supplements are limited to claims about antioxidant nutrients with established RDIs, such as vitamins C and E and beta-carotene. A supplement label may also use the term "high potency" if the product contains a vitamin or mineral present at 100 % or more of the RDI.

Appendices A and B of the FDA's *Guidance for Industry: A Food Labeling Guide* provide a list of approved nutrient content claims and the requirements for use, including the nutrient levels needed in order to use these claims [88, 89]. Only those claims, or their synonyms, that are specifically defined in the regulations may be used; all other claims are prohibited. In addition, a disclosure statement must accompany the claim if other nutrients in the product exceed certain levels [90] (see Table 1.4).

1.4.3 FTC Advertising Regulation

FTC regulates advertising for most consumer products, including dietary supplements, to ensure that consumers receive accurate and truthful information about the products they purchase. FTC has a long history of enforcement action in this area and has taken action against dietary supplement manufacturers, distributors,

Table 1.4 Claims allowed on dietary supplement labels

Type of claim	Requirements	Regulatory citation
Health claims	Petition to FDA and preapproval	*21 CFR §101.14; 21 CFR §101.70*
Qualified health claims	Petition to FDA; subject to letter of enforcement discretion	*21 CFR §101.14; 21 CFR §101.70*
Structure/function claims	FDA notification and disclaimer statement	21 CFR §101.93(f) and (g)
Nutrient content claims	FDA preapproval	21 CFR §101.13, Subpart D of part 101, and parts 105 and 107; 21 CFR §101.13(b)

marketing companies, as well as retailers. As noted previously, FDA and FTC work closely together to enforce both dietary supplement labeling and advertising restrictions. Both agencies use a similar standard to evaluate the scientific substantiation provided as support for an advertising or labeling claim.

FTC's primary role is to ensure that dietary supplement claims are truthful and not misleading, and that all claims, whether expressed or implied, are adequately substantiated prior to dissemination of the advertisement. The agency uses a "reasonable consumer" perspective when determining whether a claim is deceptive, which could be based on either a misrepresentation or the omission of pertinent information. FTC's standard for substantiation is "competent and reliable scientific evidence," which the agency uses to evaluate safety and efficacy claims for supplements, as well as claims for conventional food and drugs.

FTC's *Dietary Supplements: An Advertising Guide for Industry* [91] describes the process for identifying and interpreting advertising claims and, more importantly, it also explains how the agency applies its substantiation standard to these claims and weighs the evidence offered to support a claim. FTC also considers testimonials and endorsements to be advertising claims that should have the appropriate level of scientific evidence to support the underlying claim(s). Furthermore, testimonials should reflect what consumers will generally achieve when using the product. Otherwise, the advertiser should include a clear and conspicuous disclaimer. In 2009, FTC updated an additional guidance document on the use of endorsements and testimonials in advertising, which addresses issues such as disclosure of financial relationships and the appropriate use of customer testimonials [92].

1.4.3.1 Internet and Social Media

Advertising claims disseminated over the Internet and through social media pose unique challenges for marketers and regulators. Both FTC and FDA take the position that any required information, such as disclaimers or warnings, and any advertising disclosures required by law must be present even when there are space limitations or similar challenges. Furthermore, the information must be provided to the consumer in a way that is truthful and not misleading. FTC has a number of resources for marketers on its website specifically for online advertising and marketing compliance [93], including its *Dot Com Disclosures: Information About Online Advertising* guidance document, which advises businesses on how federal advertising law applies to advertising and sales on the Internet [94]. FTC first published this document in 2000; agency staff revised the document in 2013 to reflect the rapidly changing environment of the online world such as social media and mobile marketing [95].

1.4.3.2 Organic Claims and Environmental Benefit Claims

Dietary supplements may include claims that the ingredients or the product are "organic." Organic claims are regulated by the USDA under the National Organic Program (NOP). The NOP regulations include a definition of "organic" and provide for certification that agricultural ingredients have been produced under conditions that would meet the definition. They also include labeling standards based on the percentage of organic ingredients in food. However, food products, including dietary supplements, labeled with organic claims must comply with all applicable labeling regulations, including USDA regulations for the organic claim, FDA regulations for labeling, and FTC advertising regulations.

A recent area of interest for FTC is environmental benefit claims, which are claims about the environmental attributes of a product, package, or service in connection with the marketing or sale of such item. Examples include "non-toxic," "free of," and "recycled content" claims. FTC published its updated *Green Guides* (last updated in 1998) that provide guidance in the area of environmental claims. This topic has also garnered the interest of the advertising self-regulatory bodies such as the National Advertising Division (described in more detail below) [96].

1.4.3.3 Industry Self-Regulation and the CRN/NAD Dietary Supplement Advertising Review Program

Both FDA and FTC have robust and comprehensive regulations in the area of dietary supplements claims, as well as ample authority to enforce the law. However, self-regulation also plays an important role by enhancing marketplace surveillance of dietary supplements claims, which supports increased consumer confidence in the truth and accuracy of advertising claims for dietary supplement products, and encourages fair competition within the industry.

Since 2006, the Council for Responsible Nutrition (CRN), through its CRN Foundation, and the National Advertising Division (NAD) of the Council of Better Business Bureaus have worked together to increase monitoring of advertising for dietary supplements and functional foods [97]. NAD serves as the investigative unit of the advertising industry's system of self-regulation. The program provides companies with a forum for the prompt, voluntary discontinuance of misleading or unsubstantiated advertising claims through a cost-effective and rapid resolution process that allows advertisers the opportunity to comment, withdraw, and/or correct advertising. Companies that cooperate in this voluntary forum can also reduce the risk that their advertising is in violation of FTC law. Compliance with NAD's recommendations is voluntary; however, if an advertiser does not comply with the NAD's suggested advertising changes, if any, the claims can be referred to FTC for formal investigation.

1.5 International Regulations

This section describes regulations in the European Union, United Kingdom, Canada, and Japan. Table 1.5 also presents a comparison of regulations in these regions as well as the United States.

Table 1.5 Comparison of select countries' regulation of dietary supplements with the United States [14]

Country	Product registration	Manufacturer registration	Premarket approval	Specific good manufacturing practices	Serious adverse event reporting
United States		X (limited)		X	X
Canada	X	X Manufacturers, packagers, labelers, and importers of Natural Health Products must obtain a site license	X	X	X
United Kingdom		X	X (limited) Under EU law, novel supplements without a history of consumption in the EU prior to May, 1997		X Under EU law, firms must report any problems to local and national authorities
Japan (FOSHU products only)	X	X	X (limited) No separate GMP, firms applying for FOSHU claim must provide evidence of quality control		

1.5.1 European Union

It is no surprise that European Union regulations germane to the marketing and availability of dietary supplements (nutritional supplements) are complex and generally in a state of flux. The European Union has expanded from six original members in 1957 to 27 currently. Europe is hardly a homogenous market, with different prevailing consumer attitudes, languages, distribution channels, competitive dynamics, and established brands. Some EU member states have a long history of using these types of products, and some with a high degree of sophistication in their regulatory apparatus. The EU Directive on Food Supplements (2002/46/EC) was developed to foster harmonization of pan-European trade in this product category, and to maintain the free market availability across their borders. Public safety is paramount, so the directive further strengthened the consideration of relevant safety data, or lack thereof, when enacting provisions that would affect the entire consortium of member states. Nutritionists and allied health care professionals have also been part of the process to ensure the maintenance of adequate nutrition for both healthy and compromised individuals, with attention to both too high and too low levels of well-studied nutrients, as well as innovative newly available herbals and botanicals from within the European Union and as imports. The definition of "food supplements" within the directive is "…foodstuffs the purpose of which is to supplement the normal diet and which are concentrated sources of nutrients or other substances with a nutritional or physiological effect, alone or in combination, marketed in dose form, namely forms such as capsules, pastilles, tablets, pills, and other similar forms, sachets of powder, ampoules of liquids, drop dispensing bottles, and other similar forms of liquids and powders designed to be taken in measured small unit quantities." [98]

Recognizing the distinction between well-characterized legitimate nutrients, a "positive list" of 13 vitamins (from 32 allowable sources) and 15 minerals (from 80 allowable sources) was specified in the directive. These are contained in Annex I and II. They are not static annexes, and additional ingredients may be added after the preparation and submission of a comprehensive scientific dossier containing all relevant safety and bioavailability data, and the ingredient concerned must be thoroughly evaluated and deemed worthy to add to the Annex by the European Food Safety Authority (EFSA). Thus, the responsibility for the safety of food supplements lies with EFSA, which is also tasked with determining the safe maximum and minimum levels of vitamins and minerals [99].

Herbal and botanical products remain very popular; however, there is growing concern over their widespread availability, often via loosely monitored Internet sites. An EU directive related to traditional herbal medicines (2004/24/EC) regulates these products, with agreed-upon well-characterized ingredients and clear label language regarding uses and allowable claims [100, 101]. Furthermore, the only traditional herbal remedies permitted are those with a known efficacy, and at least 30 years of substantiated use, 15 of which must be from within the European Union.

In short, only those supplements that have been proven safe may be sold within the EU bloc without prescription; and as a category of "food," food supplements cannot be labeled with drug claims but can bear health claims and nutrition claims [102].

1.5.2 United Kingdom

Food supplements are regulated under food law. Although a great deal of food law in the United Kingdom comes from the European Union, it is important to note that the United Kingdom also has the Food Safety Act of 1990, under which all national domestic food law is made. It is the responsibility of the manufacturer, importer, or distributor to comply with the relevant legislation. The primary legislation relating to food supplements is the European Food Supplements Directive (Directive 2002/46/EC), implemented in the United Kingdom by the Food Supplements (England) Regulations of 2003. All foods, including food supplements, are also subject to the Food Labeling Regulations of 1996. The labeling regulations are in the process of being amended to conform to newer EU legislations; the Food Information for Consumers Regulation; however, this does not fully come into force until December 2014. Food supplements are generally not subject to premarket approval. For example, any supplement that either meets the guidelines established under EU law for specific vitamins and minerals, or does not include a new or genetically modified ingredient, does not require approval prior to marketing. The underpinning principles of all food law are that products have to be safe for consumption and not misleadingly labeled. This is intended to increase consumer protection and help consumers make informed choices through improved labeling requirements. In the United Kingdom, the definition of a food supplement is "…any food the purpose of which is to supplement the normal diet and which—(a) is a concentrated source if a vitamin or mineral or other substance with nutritional or physiological effect, alone or in combination; and (b) is sold in dose form" [103].

The Food Supplements Directive lists the vitamins and minerals that are permitted for use in food supplements. The Annexes have been amended several times and the safety of the substances on the list has been assessed and approved by EFSA. The Food Supplements Directive includes the intention to establish maximum and minimum levels for vitamins and minerals in food supplements. The European Commission has not yet begun this work and currently the UK industry works to the safe upper levels established by the 2003 report by the Expert Group on Vitamins and Minerals and the EU Recommended Daily Allowances (RDAs) of vitamins and minerals [104, 105].

Most direct oversight of the dietary supplement industry in the United Kingdom occurs at the local level of government (e.g., all investigations, enforcement actions, and monitoring activities such as inspections). Food supplement firms are required to register with local authorities and should detail the specific activities undertaken at each establishment as part of this process. There is no centralized information on registered firms nor is there a centralized registry of food supplement products in the United Kingdom. Although government standards for food GMP apply to food supplement manufacturing, there are no GMP guidelines specific for food supplements. Under EU law, firms are required to report any problems with food products to the local and national authorities and if the product is injurious to health, the firm must remove it from the market.

1.5.3 Canada

In 2004, under the Canadian Food and Drugs Act, the Natural Health Product (NHP) regulations were enacted. These regulations cover nutritional supplements, probiotics, traditional Chinese medicines, vitamins, minerals, herbal products, and homeopathy based upon their medicinal ingredients and intended use. The NHP Directorate (NHPD) ensures that all Canadians have ready access to natural health products that are safe, effective, and of high purity, while respecting freedom of choice and philosophical and cultural diversity. This is accomplished by implementing manufacturing quality and safety standards, while significantly relaxing the standards for product efficacy claims. Unlike the US regulatory landscape where dietary supplements are primarily monitored post-market for signs of adversity, Canada's NHPD requires products to be licensed prior to market entry. Every manufacturer, packagers, labeler, distributor, and importer of NHPs must have a "site license" from NHPD before the product(s) can be placed on the market, and this applies to domestic Canadian companies as well as all foreign entities. To obtain a site license, a firm must provide evidence of quality control procedures that meet government standards for GMP. Firms are required to report any serious adverse reactions associated with their products within 15 days and must provide information summarizing all adverse reactions, including mild or moderate events, on an annual basis. As a consequence, the foreign companies need to employ a Canadian import agent [106, 107].

Each product must have a submission number (or product license) attesting to an appropriate dossier with relevant safety and quality data and label language as having been submitted and acknowledged (or reviewed) by the premarket side of NHPD. The NHPD establishes manufacturing standards that are needed prior to the issuance of a product license. The licensure requires following GMP and in toto provides consumers and health care professionals with assurance that what is on the label is actually in the bottle. The premarket registration must be in place as demonstrated by the product being in a pending queue with a submission number, or NHPD staff review with issuance of product license number deeming that the product meets minimal standards of quality and safety and the permitted efficacy claims. This product licensing application must include detailed information about the product, ingredients, potency, intended use, and evidence supporting a product's safety and efficacy. Approved products are assigned a license number that is displayed on the product label [109, 110].

1.5.4 Japan

In Japan, there are no regulations that specifically define a food supplement. What consumers ingest is covered by either "foods" or "drugs," period. What others consider food supplements fall entirely under the food category. The general food regulations are solely applicable and any ingredient covered by the Pharmaceutical Affairs Act (i.e., a drug moiety) may not be contained in any food products. Under the Health Promotion Act, foods that comply with specifications and standards established by the Minister of Consumer Affairs Agency can be labeled with certain nutritional or health functionality and can be labeled as "Foods with Health Claims." These types of foods are grouped into two categories and are regulated based on their product claims: (1) Foods with Nutrient Function Claims (FNFC) containing well-characterized vitamins and minerals with standardized preapproved claim statements with established benefits, or (2) Foods for Specified Health Uses (FOSHU), which require government approval for safety and efficacy prior to marketing and are allowed to claim approved physiological effects on the body [110].

To date, there are over a dozen vitamins and a dozen minerals, together with over 100 botanicals/herbals that are considered bona fide foods, and thus market entry and marketing of these products is relatively straightforward. The Japanese also have a list of approved food additives, covering almost 400 synthetic and almost 500 natural origin additives. If an ingredient is not on this positive list, it is not allowed in any Japanese food or dietary supplement product for sale to the public.

FNFCs are further defined as "food products with supplement nutritional components which are likely to be deficient in the elderly and other persons who deviate from normal eating habits due to irregular lifestyle." Traditional vitamins and minerals are currently on this list, and claims are on the order of the following: "Niacin is a nutrient that helps maintain health of skin and mucus membranes," "Vitamin B12 is a nutrient that helps red blood cell formation," and so forth. Since FNFC claims are standardized and preapproved, firms do not need to notify the government prior to marketing a product using an approved FNFC claim, provided the product meets established ingredient content specifications [111].

FOSHU are defined as "foods containing functional ingredients which affect physiological and other functions of the body and which can be used to promote and maintain health or for a specific healthcare purpose." The approval comes on a case-by-case basis (i.e., each ingredient and each claim is reviewed and determined by the Japanese Ministry of Consumer Affairs), and the special dietary uses include the following: (1) formulas for pregnant or lactating women, (2) infant formulas, (3) foods for the elderly, (4) medical foods, and (5) other specialized health situations. All botanicals and herbs will go through the FOSHU channel requiring Ministry review and approval. To use a FOSHU claim on a product, a firm is required to provide the government with evidence supporting the product's physiological effect and safety prior to marketing. In addition, a firm must provide information on the company and the product it intends to market to the Japanese Ministry, as well as evidence that a quality control procedure is in place and being followed [112].

Conflict of Interest Statement All authors are employed by the Council for Responsible Nutrition (CRN). CRN, founded in 1973, is a Washington, DC-based trade association representing 100+ dietary supplement manufacturers, ingredient suppliers, and companies providing services to those manufacturers and suppliers. In addition to complying with a host of federal and state regulations governing dietary supplements in the areas of manufacturing, marketing, quality control and safety, our manufacturer and supplier members also agree to adhere to additional voluntary guidelines as well as to CRN's Code of Ethics. Visit www.crnusa.org.

References

1. Federal Food, Drug and Cosmetic Act, 52 Stat 1040 (1938), 21 USC §301 et seq.
2. Dietary Supplement Health and Education Act, 108 Stat 4325 (1994).
3. Pure Food and Drug Act, 34 Stat 768 (1906).
4. Federal Food, Drug, and Cosmetic Act, 52 Stat 1040 (1938), 21 USC §301 et seq.
5. Vitamin-Mineral Amendments, 90 Stat. 410–413 (April 22, 1976).
6. Food Safety Modernization Act, 124 Stat 3885 (2011).
7. U.S. Food and Drug Administration. FDA Food Safety Modernization Act. http://www.fda.gov/Food/GuidanceRegulation/FSMA/default.htm (Updated August 19, 2013).
8. Hutt PB. FDA statutory authority to regulate the safety of dietary supplements. *Am J Law Med.* 2005;31:155–174.
9. NIH State-of-the-Science Conference Statement on Multivitamin/Mineral Supplements and Chronic Disease Prevention. *NIH Consens State Sci Statements*. 2006;23(2):1–30.
10. 6 FR 5921 (November 22, 1941).
11. 38 FR 20,730, 20732 (August 2, 1973).
12. 1976 Proxmire Amendment, 21 USC §350 (April 22, 1976).
13. 58 FR 33690 (June 18, 1993).
14. U.S. Government Accountability Office. 2009 GAO report. FDA should take further actions to improve oversight and consumer understanding. GAO-09-250. http://www.gao.gov/products/GAO-09–250 (Published January 29, 2009).
15. Food Additives Amendment, 21 CFR §§170–186, 189 (1958).
16. Color Additives Amendment, 21 CFR §§70–82 (1960).
17. 21 CFR §§201(ff)(2)(A); 411(c)(1)(B).
18. Federal Food, Drug, and Cosmetic Act §411(c)(1)(B)(i).
19. U.S. Food and Drug Administration. *Draft Guidance for Industry: Factors that Distinguish Liquid Dietary Supplements from Beverages, Considerations Regarding Novel Ingredients, and Labeling for Beverages and Other Conventional Foods*. http://www.fda.gov/Food/GuidanceComplianceRegulatoryInformation/GuidanceDocuments/DietarySupplements/ucm196903.htm (Published December 2009).
20. 69 FR 6787 (February 11, 2004).
21. 72 FR 34751 (June 25, 2007).
22. 62 FR 49826 (September 23, 1997).
23. 65 FR 2000 (January, 6, 2000).
24. Nutrition Labeling and Eduction Act 104 Stat. 2353 (1990).
25. 62 FR 5700 (February 6, 1997).

T.C. Wallace et al., *Dietary Supplement Regulation in the United States*,
SpringerBriefs in Food, Health, and Nutrition,
DOI 10.1007/978-3-319-01502-6, © The Author(s) 2013

26. U.S. Food and Drug Administration. CFR - Code of Federal Regulations Title 21. http://www.accessdata.fda.gov/scripts/cdrh/cfdocs/cfcfr/CFRSearch.cfm?CFRPart=111 (March 13, 2013).
27. Food Safety Modernization Act, 124 Stat 3885 (2011).
28. Public Health Security and Bioterrorism Preparedness and Response Act, 116 Stat. 594 (2002).
29. U.S. Food and Drug Administration. Stimulant Potentially Dangerous to Health, FDA Warns. http://www.fda.gov/ForConsumers/ConsumerUpdates/ucm347270.htm (Published April 11, 2013).
30. U.S. Food and Drug Administration. *Guidance for Industry: Current Good Manufacturing Practice in Manufacturing, Packaging, Labeling, or Holding Operations for Dietary Supplements; Small Entity Compliance Guide*. http://www.fda.gov/Food/GuidanceRegulation/GuidanceDocumentsRegulatoryInformation/DietarySupplements/ucm238182.htm (Published December 2010).
31. 68 FR 62090 (October 31, 2003).
32. 21 CFR §111.
33. Federal Food, Drug and Cosmetic Act, §201(s).
34. Federal Food, Drug and Cosmetic Act, §402(a)(1).
35. Federal Food, Drug and Cosmetic Act, §402(f)(1)(A).
36. Federal Food, Drug and Cosmetic Act, §402(f)(1)(B); 413(a).
37. Federal Food, Drug and Cosmetic Act, §413(d).
38. Federal Food, Drug and Cosmetic Act, §413(a)(2).
39. 21 CFR §170.3.
40. Federal Food, Drug and Cosmetic Act, §403(f)(1)(C).
41. Federal Food, Drug and Cosmetic Act, §403(g)(1).
42. Food Safety Modernization Act, §102(a).
43. Food Safety Modernization Act, §102(b).
44. Food Safety Modernization Act, §206.
45. Food Safety Modernization Act, §207.
46. Federal Food, Drug and Cosmetic Act, §403.
47. Federal Food, Drug and Cosmetic Act, §413(a).
48. U.S. Food and Drug Administration. *Draft Guidance for Industry: Dietary Supplements: New Dietary Ingredient Notifications and Related Issues*. http://www.fda.gov/Food/GuidanceRegulation/GuidanceDocumentsRegulatoryInformation/DietarySupplements/ucm257563.htm (Published July 2011).
49. FDA Reacts to NDI Draft Guidance Comments. http://www.naturalproductsinsider.com/articles/2011/12/fda-reacts-to-ndi-draft-guidance-comments.aspx (Published December 20, 2011).
50. van Laack R. Senators Hatch and Harkin request that FDA withdraw Draft NDI Guidance. http://www.fdalawblog.net/fda_law_blog_hyman_phelps/2012/01/senators-hatch-and-harkin-request-that-fda-withdraw-draft-ndi-guidance.html (Published January 10, 2012).
51. Federal Food, Drug and Cosmetic Act, §402(a)(1)-(6).
52. Federal Food, Drug and Cosmetic Act, §301(ll).
53. Federal Food, Drug and Cosmetic Act, §721(a).
54. Dietary Supplement and Nonprescription Drug Consumer Protection Act, 120 Stat. 469 (2006).
55. Federal Food, Drug and Cosmetic Act, §761(a)(2).
56. Federal Food, Drug and Cosmetic Act, §761(b)(1).
57. Federal Food, Drug and Cosmetic Act, §761(c)(2).
58. Federal Food, Drug and Cosmetic Act, §761(e).
59. Memorandum of Understanding Between The Federal Trade Commission and the Food and Drug Administration. FDA-225-71-8003. http://www.fda.gov/AboutFDA/PartnershipsCollaborations/ MemorandaofUnderstandingMOUs/DomesticMOUs/ucm115791.htm (Updated March 10, 2009).

60. Federal Food, Drug, and Cosmetic Act, §403(s).
61. 21 CFR §§101.1–101.5.
62. 21 CFR §101.36(h) and (i).
63. 21 CFR §101.36(e).
64. 21 CFR §101.36(b)(2.
65. 21 CFR §101.36(b)(2)(iii);
66. 21 CFR §109(c).
67. Dietary Supplement and Nonprescription Drug Consumer Protection Act, 120 Stat 469 (2006).
68. U.S. Food and Drug Administration. *Guidance for Industry: Questions and Answers Regarding the Labeling of Dietary Supplements as Required by the Dietary Supplement and Nonprescription Drug Consumer Protection Act*. http://www.fda.gov/Food/GuidanceComplianceRegulatoryInformation/GuidanceDocuments/DietarySupplements/ucm179018.htm (Published December 2007; Revised December 2008 and September 2009). Accessed November 30, 2012.
69. 21 CFR §101.17(e)(1).
70. 21 CFR §101.9(c)(2)(ii)
71. 21 CFR §101.36(b)(2)(i).
72. U.S. Food and Drug Administration. *Guidance for Industry: A Food Labeling Guide*. http://www.fda.gov/food/guidancecomplianceregulatoryinformation/guidancedocuments/foodlabelingnutrition/foodlabelingguide/ucm064908.htm (Published October 2009).
73. U.S. Federal Trade Commission [homepage]. http://www.FTC.gov.
74. U.S. Food and Drug Administration [homepage]. http://www.FDA.gov.
75. U.S. Food and Drug Administration. Chapter VI. Claims. In: *Guidance for Industry: A Dietary Supplement Labeling Guide*. http://www.fda.gov/Food/GuidanceCompliance Regulatory Information/ GuidanceDocuments/DietarySupplements/DietarySupplement labelingguide/ucm 070613.htm (Published April 2005).
76. Federal Food, Drug, and Cosmetic Act, §403 (r)(3)(B)(i).
77. U.S. Food and Drug Administration. Appendix C: Health Claims. In: *Guidance for Industry: A Food Labeling Guide*. http://www.fda.gov/Food/GuidanceComplianceRegulatoryInformation/GuidanceDocuments/FoodLabelingNutrition/FoodLabelingGuide/ucm064919.htm (Published October 2009).
78. *Pearson v. Shalala*, 164 F.3d 650, 568 (D.C. Cir. 1999).
79. U.S. Food and Drug Administration. *Guidance for Industry: Interim Procedures for Qualified Health Claims in the Labeling of Conventional Human Food and Human Dietary Supplements*. http://www.fda.gov/Food/GuidanceComplianceRegulatoryInformation/GuidanceDocuments/FoodLabelingNutrition/ucm053832.htm (Published July 2003).
80. U.S. Food and Drug Administration. *Guidance for Industry: FDA's Implementation of "Qualified Health Claims": Questions and Answers; Final Guidance*. http://www.fda.gov/food/guidancecomplianceregulatoryinformation/guidancedocuments/foodlabelingnutrition/ucm053843.htm (Published May 12, 2006).
81. U.S. Food and Drug Administration. *Guidance for Industry: Evidence-Based Review System for the Scientific Evaluation of Health Claims—Final*. http://www.fda.gov/Food/GuidanceComplianceRegulatoryInformation/GuidanceDocuments/FoodLabelingNutrition/ucm073332.htm (Published January 2009).
82. U.S. Food and Drug Administration. Appendix D: Qualified Health Claims. In: *Guidance for Industry: A Food Labeling Guide*. http://www.fda.gov/Food/GuidanceComplianceRegulatoryInformation/GuidanceDocuments/FoodLabelingNutrition/FoodLabelingGuide/ucm064923.htm (Published October 2009).
83. U.S. Food and Drug Administration. Claims that can be made for conventional foods and dietary supplements. http://www.fda.gov/food/labelingnutrition/labelclaims/ucm111447.htm (Published September 2003).
84. Federal Food, Drug, and Cosmetic Act, §403(r)(6).

85. U.S. Food and Drug Administration. *Guidance for Industry: Structure/Function Claims, Small Entity Compliance Guide*. http://www.fda.gov/food/guidancecomplianceregulatory information/ guidancedocuments/dietarysupplements/ucm103340 (Published January 9, 2002).
86. Regulations on Statements Made for Dietary Supplements Concerning the Effect of the Product on the Structure or Function of the Body. 65 FR 1000 (January 6, 2000).
87. U.S. Food and Drug Administration. *Guidance for Industry: Substantiation for Dietary Supplement Claims Made Under Section 403(r)(6) of the Federal Food, Drug, and Cosmetic Act*. http://www.fda.gov/Food/GuidanceComplianceRegulatoryInformation/ GuidanceDocuments/DietarySupplements/ucm073200.htm (Published December 2008).
88. U.S. Food and Drug Administration. Appendix A: Definitions of Nutrient Content Claims. In: *Guidance for Industry: A Food Labeling Guide*. http://www.fda.gov/Food/Guidance ComplianceRegulatoryInformation/GuidanceDocuments/FoodLabelingNutrition/ FoodLabelingGuide/ucm064911.htm (Published October 2009).
89. U.S. Food and Drug Administration. Appendix B: Additional Requirements for Nutrient Content Claims. In: *Guidance for Industry: A Food Labeling Guide*. http://www.fda.gov/ Food/GuidanceComplianceRegulatoryInformation/GuidanceDocuments/ FoodLabelingNutrition/FoodLabelingGuide/ucm064916.htm (Published October 2009).
90. 21 CFR §101.13(h).
91. U.S. Federal Trade Commission. *Dietary Supplements: An Advertising Guide for Industry*. http://business.ftc.gov/documents/bus09-dietary-supplements-advertising-guide-industry#c (Published April 2001).
92. 16 CFR Part 255: Guides Concerning the Use of Endorsements and Testimonials in Advertising: Notice Announcing Adoption of Revised Guides. http://www.ftc.gov/opa/ 2009/10/endortest.shtm (Published October 5, 2009).
93. U.S. Federal Trade Commission. Online advertising and marketing [webpage]. http://business. ftc.gov/advertising-and-marketing/online-advertising-and-marketing.
94. U.S. Federal Trade Commission. *Dot Com Disclosures: Information About Online Advertising*. http://www.business.ftc.gov/sites/default/files/pdf/bus41-dot-com-disclosures-information-about-online-advertising.pdf (Published May 2000).
95. U.S. Federal Trade Commission. *Dot Com Disclosures: How to Make Effective Disclosures in Digital Advertising*. http://www.ftc.gov/os/2013/03/130312dotcomdisclosures.pdf (Published March 12, 2013).
96. U.S. Federal Trade Commission. *Environmental Claims: Summary of the Green Guides*. http://business.ftc.gov/documents/environmental-claims-summary-green-guides (Published October 2012).
97. Council for Responsible Nutrition Foundation/National Advertising Division. Industry Self-Regulatory Program [webpage]. http://www.crnusa.org/NAD/.
98. European Commission. Directive 2002/46/EC of the European Parliament and of the Council of 10 June 2002 on the approximation of the laws of Member States relating to food supplements. *Official J EC* 2003;L183:51–57.
99. Scientific Committee on Food. *Guidance on Submissions for Safety Evaluation of Sources of Nutrients or of Other Ingredients Proposed for Use in the Manufacture of Foods*. SCF/CS/ ADD/NUT/21 Final. http://ec.europa.eu/food/fs/sc/scf/out100_en.pdf (Published 2001).
100. European Commission. Directive 2003/89/EC of the European Parliament and of the Council of 10 November 2003 amending Directive 2000/13/EC as regards identification of the ingredients present in foodstuffs. http://ec.europa.eu/food/food/labellingnutrition/foodlabelling/ fl_com2003–89_en.pdf. (Published 2003).
101. European Commission. Directive 2004/24/EC of the European Parliament and of the Council of 31 March 2004 amending, as regards traditional herbal medicinal products, Directive 2001/83/EC on the Community code relating to medicinal products for human use. *Official J EC* 2004;L136:85–90.
102. Eberhardie C. Nutritional supplements and the EU: is anyone happy? *Proc Nutr Soc*. 2007;66:508–511.

103. United Kingdom. The Food Supplements (England) Regulations of 2003. http://www.legislation. gov.uk/uksi/2003/1387/contents/made (Published 2003).
104. U.K. Food Standards Agency. Joint Food Safety and Standards Group; Expert Group on Vitamins and Minerals http://archive.food.gov.uk/dept_health/archive/evm.htm (Published 2000).
105. European Commission. Commission Regulation No 1161/2011 of 14 November 2011 amending Directive 2002/46/EC of the European Parliament and of the Council, Regulation (EC) No 1925/2006 of the European Parliament and of the Council and Commission Regulation (EC) No 953/2009 as regards the lists of mineral substances that can be added to foods. http://eur-lex.europa.eu/LexUriServ/LexUriServ.do?uri=OJ:L:2011:296:0029:0030:EN:PDF (Published 2011).
106. Wojewnik P. The Canadian Marketplace and Regulations for Dietary Supplements Articles, Topics, Good Manufacturing Practices, International Quality Control. http://www.naturalproductsinsider.com/articles/2010/01/canadian-supplement-market-and-regs.aspx (Published 2010).
107. Gavura S. Supplement Regulation: Be Careful What You Wish For Herbs, Politics and Regulation, Science and Medicine. http://www.sciencebasedmedicine.org/index.php/supplement-regulation-be-careful-what-you-wish-for/ (Published 2013)
108. Health Canada. Natural Health Products Regulations SOR/2003–196. http://laws-lois.justice.gc.ca/eng/regulations/SOR-2003–196/FullText.html (Published 2003).
109. Health Canada. Natural Health Products (NHPs). http://www.hc-sc.gc.ca/dhp-mps/prodnatur/ index-eng.php (Published 2013).
110. Shiozaki AH, Ueda Y. Japan. In: *Differences and Similarities in Food Supplement Regulations*. Houston, TX: Lex Mundi Ltd.; 2010.
111. Anonymous. Dietary Supplements in Japan in International Business Strategies. http://www.fda.gov/ohrms/dockets/dockets/95s0316/95s-0316-rpt0275–23-Inter-Business-Strategies-vol211.pdf (Published 2001).
112. Japanese Ministry of Health, Labour and Welfare. Food Safety Information: Food Additives. http://www.mhlw.go.jp/english/topics/foodsafety/foodadditives/ (2013).
113. Soller WR, Bayne HJ, Shaheen C. The regulated dietary supplements industry: Myths of an unregulated industry dispelled. *HerbalGram*. 2012;93:42–57.

FSC
www.fsc.org
MIX
Papier aus verantwortungsvollen Quellen
Paper from responsible sources
FSC® C105338

If you have any concerns about our products,
you can contact us on
ProductSafety@springernature.com

In case Publisher is established outside the EU,
the EU authorized representative is:
Springer Nature Customer Service Center GmbH
Europaplatz 3, 69115 Heidelberg, Germany

Printed by Libri Plureos GmbH
in Hamburg, Germany